The Concise Handbook of
ASTRONOMY

The Concise Handbook of ASTRONOMY

IAN RIDPATH

GALLERY BOOKS
An Imprint of W. H. Smith Publishers Inc.
112 Madison Avenue
New York City 10016

This book was devised and produced by
Multimedia Publications (UK) Ltd.

Editors: Jeff Groman, Andie Oppenheimer
Production: Karen Bromley
Design: John Youé & Associates Limited
Picture Research: Mirco Decet

Compilation Copyright © Multimedia Publications (UK) Ltd 1986
Text Copyright © Ian Ridpath 1986

All rights reserved. No part of this book may be
reproduced or transmitted in any form or by any
means, electronic or mechanical, including photocopying
and recording, or by any information storage retrieval
system, without permission in writing from the
publisher and the copyright holders.

First published in the United States of America 1986 by
Gallery Books, an imprint of W. H. Smith Publishers Inc.,
112 Madison Avenue, New York, NY 10016
ISBN 0 8317 1766-1

Typeset by Tradespools Limited
Origination by Peninsular Repro Service Limited
Printed by Cayfosa, Barcelona, Spain

CONTENTS

Astronomy and Astronomers	6
Eyes on the Universe	16
Exploring the Solar System	30
Sky Wanderers	50
Furnaces in the Skies	54
Exotic Objects	66
Galaxies and Beyond	74
The New Astronomy	90
Index	96

Front cover: Giant radio dish making up a section of the Very Large Array, New Mexico.

Back cover and page 1: An observatory in the Matra Mountains in Hungary.

Pages 2 and 3: The Horsehead Nebula silhouetted against brightly glowing gas in the constellation Orion.

These pages: The 32.9-m (110-ft) tower at Kitt Peak Observatory.

ASTRONOMY AND ASTRONOMERS

Have you ever gazed into the night sky and wondered what the stars are, how far away they are, and how the vast Universe in which we live came into being? Astronomy, the scientific study of the heavens, attempts to answer these fascinating and fundamental questions. With the help of giant telescopes on Earth and satellites in space, astronomers have discovered the true nature of stars and planets, and have pieced together much of the history of the known Universe. They can now make limited predictions about what may happen in the future: how the Sun and Earth may die, and what will be the eventual fate of the Universe.

Astronomy is both the oldest and the youngest of the sciences. Records of events in the sky go back to the birth of writing, thousands of years BC, while a growing number of people believe that prehistoric monuments such as Stonehenge, England, embody an advanced astronomical knowledge because they are aligned on the rising and setting points of the Sun and Moon. Now space-age technology has given astronomers their first clear views of the

Above: The 76-m (250-ft) radio telescope at Jodrell Bank, England, operated by the University of Manchester. Radio waves from space are reflected to a receiver at the center of the dish.

Left: Stonehenge, the prehistoric stone monument on Salisbury Plain in England, may have been used as an observatory to follow the motions of the Sun and Moon.

heavens, unobscured by the dense and turbulent blanket of the Earth's atmosphere, and even on-the-spot views of our close celestial neighbors, the Moon and planets.

While the impressive technological aids of the modern professional astronomer have revolutionized our knowledge of the space around us, for countless amateur astronomers who scan the sky with binoculars or telescope each clear night, the basic lure of the stars remains – the same lure that has always turned human eyes and imagination skywards.

Star patterns

Although the stars at first sight seem to be countless, on even a dark, clear night there are no more than about 2,000 stars visible to the naked eye at any one time. Thousands of years ago, imaginative astronomers grouped the stars into patterns known as constellations, representing mythical gods and heroes. Astronomers continue this convenient tradition today: there are a total of 88 constellations covering the entire sky.

The stars visible each night change with the seasons as the Earth orbits the Sun. For instance, the magnificent constellation of Orion, the hunter, is easily visible in the early evening during December, January, and February, but by June it is invisible from Earth, having passed behind the Sun. The aspect of the skies also changes with one's latitude on Earth – the altitude of the north or south celestial pole above the horizon depends on how far north or south of the equator you are.

To begin an interest in astronomy you need no optical aids at all, just your eyes. Familiarizing yourself with the appearance of the heavens is the first important step. Dedicated amateurs, using nothing more elaborate than binoculars, cameras, or their eyes alone, have been able to feed professional astronomers with important information on stars that erupt into prominence or fluctuate in brightness. Amateurs also carefully observe the tiny particles of interplanetary debris that burn up in the Earth's atmosphere to become the brilliant streaks known as meteors. Others help to track satellites as they orbit the Earth, either by simple visual observations or by picking up the transmissions from the satellites with short-wave radio. For example, in 1966 a group of radio trackers at Kettering School, England, discovered that Soviet satellites were being launched from a secret new site at Plesetsk in northern Russia. Eleven years later the same group predicted the crash of a Soviet nuclear-powered satellite, Cosmos 954, which subsequently scattered radioactive debris over Canada.

Ancient views

We now know that stars are glowing balls of gas similar to our own Sun, only much farther away – so far, in fact, that their light takes years to reach us. But to a casual observer the stars appear to lie on a dome encompassing the Earth and, until a few centuries ago, most people believed that this really was the case. They thought that the dome of stars rotated around the Earth once each day, along with the Sun, Moon, and planets. This view, known as the geocentric theory, assumed that the Earth was the center of the Universe. Astronomers of the time did not know the true nature of the stars, nor did they realize that planets are non-luminous bodies like the Earth that shine by reflecting sunlight. These facts did not emerge until after the invention of the telescope in the 17th century.

Ancient notions of the Universe, before the rise of Greek civilization, were admittedly fanciful. The people of the Middle East, thousands of years before Christ, pictured the Earth as a flat slab surrounded by water. To the ancient Egyptians, the sky was the star-studded body of the goddess Nut; the Sun-god Ra sailed in a boat across the sky each day, while the planets voyaged out in their own boats at night. In the Hindu view, our world was carried on the backs of four elephants.

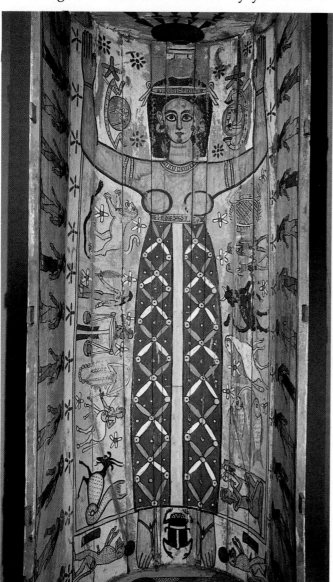

Left: Nut, the Egyptian goddess of the sky, shown surrounded by signs of the zodiac on the inside of a wooden coffin from Thebes.

Right: One of 12 stone sundials, called Rasivalaya Yantras, used for observing the Sun in different signs of the zodiac, and part of an early 18th-century observatory of stone instruments, at Jaipur, India.

The first astronomers

Perhaps the first true astronomer was the Greek Anaximander (c.610–c.546 BC) who realized that the Earth's surface is curved, not flat; he pictured the Earth as a short cylinder suspended at the centre of the Universe. The Greek mathematician Pythagoras, in the 6th century BC, made the first recorded suggestion that the Earth is a sphere, an idea that would have seemed quite sensible to navigators and other travellers who observed how the altitude of the pole star changed as they moved north or south. Pythagoras believed that the Sun, Moon, and the five wandering 'stars' known as planets moved around the Earth on crystalline spheres, each of which made its own heavenly music as it turned – the music of the spheres.

Pythagoras's concern for musical harmony seems to have outweighed his regard for strict astronomical accuracy, but his concept of a spherical Earth gained widespread acceptance. In one of the great achievements of the ancient world, the Greek scientist Eratosthenes, in about 240 BC, calculated the circumference of the Earth to be 40,000 km (25,000 miles), to an accuracy of a few per cent. He did so by observing the different altitude of the Sun, caused by the Earth's curvature, seen from two places in Egypt of known distance apart. It is surprising that Columbus, 1,700 years later, should have believed that the Earth was only about half its actual size. This error is one reason why Columbus had trouble finding sponsorship for his proposed transatlantic voyage. Those to whom he first turned probably knew the size of the Earth better than he did.

In the 3rd century BC, the Greek astronomer Aristarchus estimated the relative distances of the Sun and Moon, and calculated their sizes in relation to the Earth. Aristarchus was nearly right in his estimate that the Moon is one-third the diameter of the Earth – actually it is 3,476 km (2,160 miles) in diameter against the Earth's 12,756 km (7,926 miles). But his estimate that the Sun is seven times the Earth's size and lies twenty times farther away than the Moon was far short of the mark – the Sun is actually 109 Earth diameters in width, and lies at a distance of 150 million km (93 million miles) as against the Moon's distance of 384,400 km (239,000 miles).

Despite this serious underestimate, Aristarchus had at least shown that the Sun was clearly bigger than the Earth, which was his justification for proposing that the Sun was the center of the Universe, with all other bodies revolving around it rather than around the Earth. This so-called heliocentric or Sun-centered theory did not gain favor among Aristarchus's contemporaries, and was not revived until the time of Copernicus, 1,800 years later.

Scientists preferred instead to follow the word of the Greek philosopher Aristotle (384–322 BC), who denied that the Earth moved in space or even that it turned upon its axis. He asserted that the Earth was immovably fixed at the center of the Universe, although he did agree that it was a sphere. To Aristotle, the heavens were perfect and unchanging, and like Pythagoras before him he held that the Sun, Moon, and planets moved around the Earth on crystalline spheres. Unfortunately, detailed observations had shown that the movement of the celestial bodies was not uniform, so Aristotle devised a complex system involving a total of 55 spheres rotating at different speeds in an attempt to account for these irregular motions. But this cumbersome and unconvincing scheme was later abandoned.

Hipparchus the pioneer

The greatest of all Greek astronomers was Hipparchus who, in the 2nd century BC, made a catalogue of 850 stars which was still in use by astronomers in the Middle Ages. Hipparchus divided the stars in his catalogue into six classes of brightness, called magnitudes. First-magnitude stars were the brightest, while sixth-magnitude stars were the faintest visible to the naked eye. Astronomers use a refined version of this magnitude system today.

Hipparchus observed the motions of the Sun and Moon as accurately as his simple sighting instruments would allow. His results led to improved accuracy in the prediction of eclipses, which happen when the Moon blocks off the Sun's light from Earth (a solar eclipse) or when it passes into the Earth's shadow (a lunar eclipse). Hipparchus abandoned Aristotle's clumsy system of multiple spheres and showed instead that the Sun's motion could be accounted for by a circle with the Earth not quite at the center. He was, however, unable to account accurately for the motion of the Moon by such a simple system, and he left the even more difficult problem of the motion of the planets untouched.

Ptolemy's encyclopedia

The final landmark in Greek astronomy was Ptolemy (c.100–c.178 AD), who wrote a major astronomical encyclopedia now known by its Arabic name of the Almagest (the greatest) – much of it based on the work of Hipparchus. In the Almagest, Ptolemy attempted to summarize all previous Greek astronomy and to produce a final description of the motions of celestial bodies around the Earth. He adopted the idea that the basic orbits of the Moon and planets were circles whose centers were offset from the Earth. Like all Greek astronomers, Ptolemy embraced the erroneous assumption that the heavens were perfect, and that only the 'perfect' shape of the sphere or the circle was good enough for the motions of the celestial bodies.

For nearly 1,500 years after Ptolemy, astronomy in

Above: Claudius Ptolemy, who theorized that the Earth was the center of the Universe, around which all else orbited.

Europe entered a period of total eclipse – the Dark Ages. The knowledge of the Greeks passed into Arab hands, where it was preserved and eventually passed back to Europe through the Arab conquest of Spain. Arab astronomy bequeathed us the popular names of many stars, such as Aldebaran, Altair, and Algol.

Copernicus and the Renaissance

Astronomy was eventually shaken out of its dormancy during the Renaissance by a Polish monk, Nicolaus Copernicus (1473–1543). By this time it had become clear that there were serious discrepancies between the observed positions of the planets and those predicted by the geocentric theory of Ptolemy. Against the tide of accepted astronomical thought, Copernicus turned back to the Sun-centered or heliocentric theory originated by Aristarchus. He believed that in doing so he could produce a simpler and more accurate description of the movements of the heavenly bodies. On this theory, a planet's speed of movement in the sky depended on its distance from the Sun, from quicksilver Mercury at the nearest out to sluggish Saturn, the farthest planet then known. If Mercury and Venus orbited closer to the Sun than the Earth, this would explain why they never appeared far from the Sun in the sky. The otherwise puzzling backward loops occasionally taken by planets such as Mars could be readily explained by the Earth catching up and overtaking more distant planets on their larger and slower orbits around the Sun.

Copernicus published his theory in the year he died, 1543, in a book entitled *On the Revolutions of the Celestial Spheres*. Unfortunately, the theory suffered one major flaw: it still described the orbits of the planets in terms of complex combinations of circles and epicycles. In many ways, therefore, it seemed little improvement on the theory of Ptolemy. One criticism levelled against the heliocentric theory was that if the Earth really did move then the positions of the stars should seem to change during the year, in the same way that a tree shifts against the background as one moves around a field. Copernicus answered this objection by arguing (correctly as we now know) that the stars must be exceptionally distant compared with the Sun; but at the time his argument seemed unconvincing.

Tycho Brahe
A new attack on the problem of planetary motion was made by the Danish astronomer Tycho Brahe (1546–1601), the last great observer of the pre-telescopic era. Tycho realized that all existing tables of the planetary motions were inaccurate, even those of Copernicus – so he set out to make precise observations from which new theories of planetary motion could be derived. Tycho's painstaking observations overthrew many cherished beliefs about the heavens. He proved that a new star which flared up temporarily in the constellation of Cassiopeia in 1572 lay far off in space, thus contradicting the Greek dogma that the heavens were unchanging and that all such transient events originated in the Earth's atmosphere. (The remains of this exploded star, known as Tycho's supernova, are still detectable by astronomers today.) Five years later Tycho showed that the bright comet of 1577 moved among the orbits of the planets, thus finally shattering any remaining faith in the Greek concept of crystalline spheres.

Tycho never accepted the Copernican theory; instead he developed his own compromise theory in which the planets orbited the Sun while the Sun and stars orbited the Earth. He took as his assistant a young German mathematician, Johannes Kepler (1571–1630) to whom he bequeathed his observations in the hope that Kepler might prove the Tychonic view of the heavens. After six years of calculation, Kepler discovered the truth about the planets' orbits – but rather than supporting Tycho, his results established the theory of Copernicus.

Kepler's laws of planetary motion
Kepler found that the planets travel around the Sun not in complex combinations of circular motions but along simple curves called ellipses. The planets speed up and slow down along their elliptical orbits as they move nearer to or farther from the Sun. Kepler later discovered a formula that links a planet's distance from the Sun with the time it takes to complete its orbit. Kepler's laws of planetary motion are at the heart of our modern understanding of the Solar System.

Even as Kepler published his discovery about the shape of planetary orbits in 1609, other events were happening that would lead to the triumphant acceptance of the Copernican theory. In that same year, the Italian scientist Galileo Galilei (1564–1642) heard of the invention of the telescope and decided to construct one for himself. (A Dutch optician, Hans Lippershey, c.1570–1619, is usually credited with the invention of the telescope, although other people experimenting with lenses almost certainly discovered the principle before he did.)

The telescope Galileo built is of the type known as a refractor, which uses a lens called an object glass to collect light and to focus it so that it can be magnified by another lens known as the eyepiece. Galileo's best telescope had an object glass only 44 millimeters (1.75 inches) in aperture and magnified objects 33 times. Optically it was crude – a modern pair of binoculars will show the sky as clearly as Galileo saw it – but what he was able to discern with this relatively primitive instrument created a revolution in scientific theories.

Right: Two of Galileo's telescopes, which he referred to as Optic Tubes, on display in the Museum of Physics and Natural History in Florence.

Left: Galileo Galilei, the Italian scientist whose observations through his home-made telescopes in 1610 confirmed Copernicus' theory that the Earth orbited the Sun like the other planets.

Above: Sir Isaac Newton was one of the greatest scientists in history. His laws of gravity explained why the planets orbited the Sun.

Left: Some early refractors had no tube; in this 'aerial telescope' the main lens (the object glass) is mounted on a tall pole, while the observer looks through an eyepiece attached to it by a long thread.

Galileo's revolution

Galileo found that the surface of the Moon was pockmarked with craters – further proof that the objects in the heavens were not perfect as the Greeks had claimed. Turning to the Milky Way, the hazy band of light that crosses the sky, Galileo could see through his telescope that it was made up of countless stars too faint to be individually visible to the naked eye. This was an indication that the Universe was infinitely vaster than the traditional views of astronomy had acknowledged. Since the stars remained as points of light in his telescope, and could not be seen as disks like the planets, the stars must indeed be very far away, as Copernicus had maintained. Most significantly of all, Galileo saw for the first time that the planet Venus went through a cycle of phases similar to those of the Moon. And the only way that Venus could show changing phases was if it orbited the Sun. Here was observational verification of the theory of Copernicus. Further direct support for Copernicus came from Galileo's discovery of four bright moons going around Jupiter, which he likened to a scaled-down version of the planets orbiting the Sun.

The new view of the heavens given by the telescope, coming hard on the heels of Kepler's theoretical breakthrough, meant that the old Earth-centered view of the heavens was gone forever. Even the efforts of the Catholic church, which put Galileo on trial and made him publicly reject the heliocentric theory, could not prevent the overthrow of traditional concepts of astronomy and

Above: The first reflecting telescope had a metal mirror 5 cm (2 inches) in diameter and was built in 1668 by Isaac Newton.

mankind's place in the Universe. By 1687, when Sir Isaac Newton (1642–1727) published his theory of gravity which explained in physical terms why planets orbited the Sun as they do, there could no longer be any doubt that the heliocentric theory was correct.

Newton advanced astronomy in another way in 1668 when he built a telescope of the kind known as a reflector, which uses mirrors instead of lenses to collect and focus light. Reflectors soon caught on, both in Newton's design and in other designs originated by the Scot James Gregory, who in 1663 had been the first to propose the principle of the reflecting telescope – although he never built one.

EYES ON THE UNIVERSE

In 1781 the English astronomer William Herschel (1738–1822), using a reflector of 15 cm (6 inches) aperture, discovered a previously unknown planet, Uranus. Until that time the Solar System had been thought to end at Saturn, which is the farthest planet from the Sun visible to the naked eye. Herschel's discovery of Uranus doubled the size of the known Solar System and awakened astronomers to the possibility of still more planets. In 1846 and 1930 two more planets, Neptune and Pluto, were discovered, both as the result of deliberate searches.

Using reflectors up to 122 cm (48 inches) aperture, Herschel went on to make the most thorough surveys of the starry sky. His studies convinced him that the stars were not arranged uniformly in space, but that they were concentrated into an irregular lens shape, with the Sun at or near the hub; where we look along the main thickness of stars we see the Milky Way. Although the Earth had been displaced

Below: Curving star trails above the domes of the 2.3-m (90-inch) Steward reflector (left) and the 4-m (158-inch) Mayall reflector on Kitt Peak, Arizona.

Above: Artist's impression of the remote planet Uranus, discovered by William Herschel in 1781. In 1977 astronomers discovered that the planet is encircled by several thin, faint rings.

Right: The Moon, seen through an amateur's telescope. A curving chain of three large craters, Theophilus, Cyrillus and Catharina, lies on the edge of the dark lowland Mare Nectaris. Theophilus, the largest of the craters shown here, is 100 km (60 miles) across.

from its position as center of the Universe, the Sun was now thought to be near the center of a Galaxy of stars which was presumed to constitute the entire Universe.

Early telescope mirrors

Mirrors of that early era were made of a shiny metal alloy of copper and tin which tarnished and had to be regularly repolished. During the 19th century, telescope mirrors began to be made as they are today – of glass coated with a reflective layer of silver or aluminum, which provides a much brighter image than the old-fashioned metal mirrors. Whereas lenses must be made of clear, flawless glass accurately shaped and polished on both sides, mirrors are not transparent and are figured on one surface only. Therefore they are much cheaper to

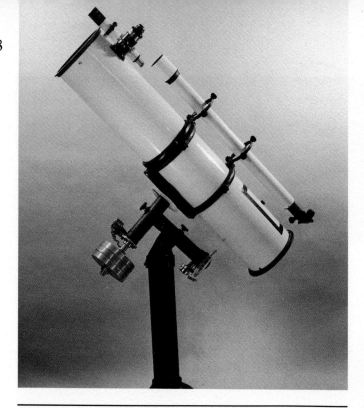

Above: A large reflecting telescope of Newtonian design on an equatorial mounting – the type of instrument many amateur astronomers aspire to own.

Below: Lick Observatory, on Mount Hamilton near San Jose, is operated by the University of California. At center is the dome of its 3-m (120-inch) reflector.

Far right: The stardust trail of the Milky Way is densest towards the center of our Galaxy, in the constellation Sagittarius. Pink patches are glowing clouds of gas.

make than lenses and, because a mirror can be supported underneath to stop it from sagging out of shape, they can be made much larger than lenses. This is important because a telescope's most vital statistic is its aperture, not its magnification. The wider the aperture, the more light it collects and thus the fainter the objects it can see. Today, astronomers use their telescopes mostly as giant telephoto lenses to take long-exposure photographs of faint objects in the sky, or to collect light for analysis by subsidiary instruments such as spectroscopes. Spectroscopic analysis of an object's light can reveal an amazing amount of information, particularly what stars are made of. Spectroscopy has become one of the most important techniques of the astronomer.

The advantages of cheapness and potential size eventually proved decisive for the popularity of reflectors. The largest refractors ever made were the 91-cm (36-inch) of Lick Observatory and the 1-m (40-inch) of Yerkes Observatory, which were completed by the American firm of Alvan Clark and Sons in 1888 and 1897 respectively. Since then, all the world's largest telescopes have been reflectors.

Giant eyes on the Universe

The modern era of large reflectors sited in the clear air of high mountains began in 1908 when the American astronomer George Ellery Hale (1868–1938) established the 1.5-m (60-inch) reflector on top of Mount Wilson in California, 1,742 metres (5,700 ft) high. With this telescope, the American astronomer Harlow Shapley (1885–1972) made a

Left: Kitt Peak National Observatory, at an altitude of 2,000 m (6,700 ft) in the Quinlan Mountains outside Tucson, Arizona, has the world's greatest concentration of telescopes.

Above: The world's highest astronomical observatory is at an altitude of 4,200 m (13,800 ft) on Mauna Kea, Hawaii, where the 3.6-m (142-inch) Canada-France-Hawaii telescope is situated.

discovery as momentous in its own way as the theory of Copernicus. Shapley found that our Sun lies not at the center of our Galaxy but about two-thirds of the way to the edge. Our Galaxy, which contains at least 100,000 million stars, is 100,000 light years in diameter – that is to say, a beam of light travelling at 300,000 km (186,000 miles) per second takes 100,000 years to cross it.

Hale went on to organize the construction of an even bigger telescope, the Mount Wilson 2.5-m (100-inch) reflector opened in 1917, which provided more revolutionary discoveries. With this giant eye, Edwin Hubble (1889–1953) discovered that other galaxies exist outside our own, and that they seem to be receding from us at speeds that increase with their distance, as though the entire Universe is expanding – the starting point for modern theories of cosmology.

Present-day reflectors

There are now many reflectors throughout the world of 2.5 meters aperture and above. The most famous is the 5-m (200-inch) Hale reflector, situated 1,700 meters (5,600 ft) high on Palomar Mountain, California, opened in 1948. This was the world's largest reflector until the opening in 1976 of the Soviet 6-m (236-inch) reflector at Zelenchukskaya in the Caucasus mountains. Other large reflectors are at Kitt Peak, Arizona, and Cerro Tololo, Chile (both 4 meters, 158 inches); at the Anglo-Australian Observatory, New South Wales (3.9 meters, 153 inches); the European Southern Observatory in Chile, and the Canada–France–Hawaii telescope on Mauna Kea, Hawaii, (both 3.6 meters, 142 inches).

An important innovation has been the introduction of reflectors which use not one large mirror but several smaller ones. The first example was the Multiple Mirror Telescope on Mount Hopkins near Tucson, Arizona, which began operation in 1979. Its six mirrors of 1.8-m (72 inches) aperture equal the light-gathering power of a 4.5-m (176-inch) mirror, making it the third-largest telescope in the world. The advantage of multi-mirror telescopes is that they are cheaper to build than telescopes with single large mirrors. Multiple-mirror designs will make it possible in the future to build telescopes far larger than any in operation today.

Above: For astrophotography, a camera body can be attached directly to the telescope, with or without the eyepiece in place.

Choosing telescopes and binoculars

To begin observing the sky, you do not need a telescope. Far better to start with a simple pair of binoculars, which will bring into view objects that are fainter than visible to the naked eye. Binoculars will be of value even if you later progress to a telescope. They also have the advantage of being inexpensive.

Binoculars are marked with figures such as 8×30, 7×50 and 10×50. The first figure gives the magnification, that is, how much bigger the object appears when viewed through the binoculars than when it is seen with the naked eye. For instance, if the binoculars magnify eight times, the object under view will appear eight times larger (and hence eight times nearer). This will allow you to see craters on the Moon, for example. However, for most astronomical purposes the magnification is less important than the amount of light the binoculars gather.

The bigger the lenses at the front of the binoculars, the more light they collect and hence the fainter the objects that can be seen. The width, or aperture, of the front lenses is given by the second figure.

Right: Three main designs of telescope. In the refractor, at top left, a lens known as the object glass collects and focuses light into an eyepiece, which magnifies the resulting image. Top right is a Schmidt telescope, for making wide-field photographs of the sky. The Schmidt uses both a lens at the front of the tube and a mirror at the rear to collect and focus light onto a photographic plate placed inside the tube. At bottom is a reflector, the most common design. A large main mirror collects light, which can be focused to a point inside the tube (known as the prime focus, insert), or it can be reflected back through a hole in the main mirror to the Cassegrain focus.

Binoculars usually have apertures of between 30 and 50 mm and have the advantage of being compact and portable. A comfortable way of scanning the skies with binoculars is lying in a deckchair or sun bed. They are ideal for studying objects such as star clusters, variable stars, nebulae, galaxies and comets.

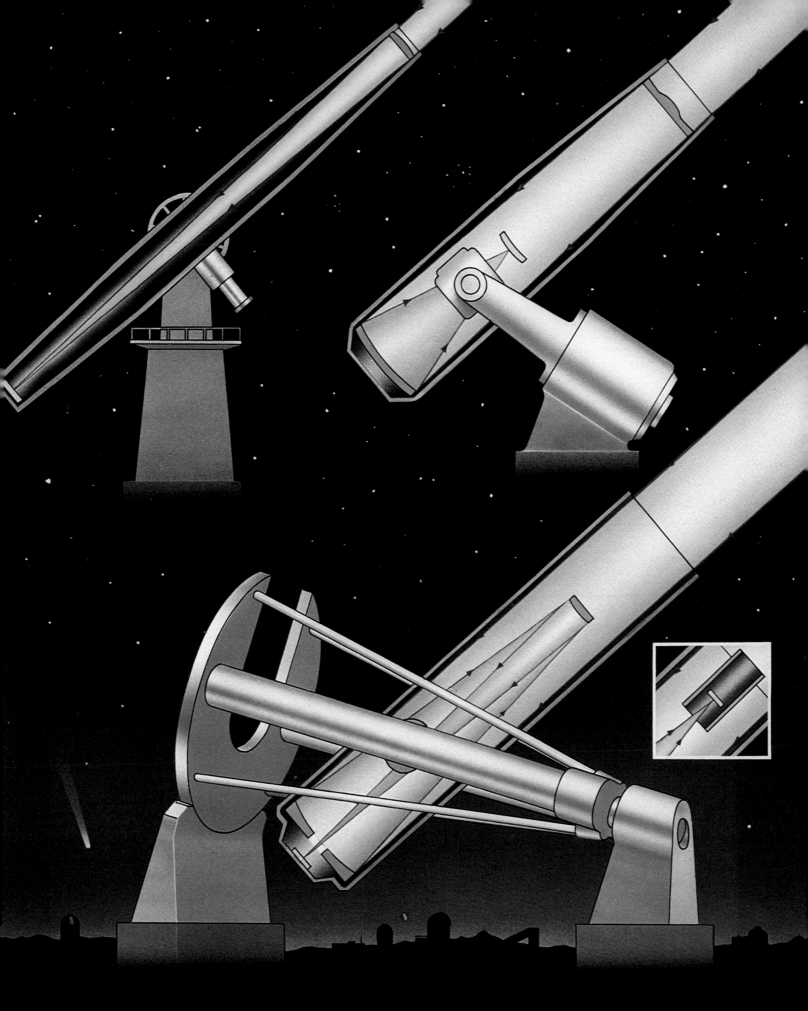

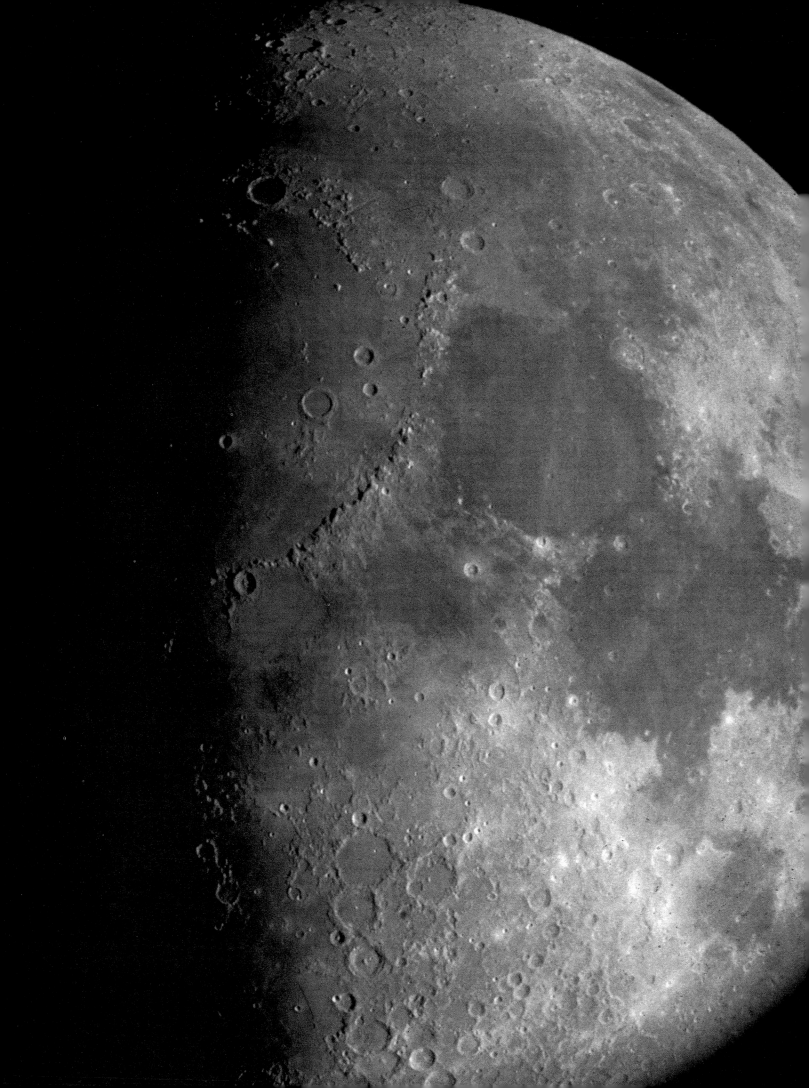

Telescopes

If you want to look at planets or close double stars, however, a telescope is necessary. Telescopes are described by the aperture of the lens or mirror which collects light – for example, a 75-mm (3-inch) refractor or a 150-mm (6-inch) reflector. Telescopes have interchangeable eyepieces which provide a range of magnifications – a low power for large objects and higher powers for homing in on the fine detail of the Moon and planets. However do not be tempted to use powers that are too high or else the image will become faint and indistinct. Remember also that the atmosphere will blur the view when you use high magnifications. For these reasons, it is a good rule to use no more than a magnification of 20 times for every 10 mm of aperture (50 times per inch).

Small telescopes are widely available in stores selling optical equipment, but they are often no more than toys. A good telescope is a high-quality precision instrument, like a large telephoto lens for a camera, and is therefore expensive. Remember that an astronomical telescope turns the image upside-down, so such a telescope will be of limited use for terrestrial viewing without additional eyepieces that turn the image the right way up.

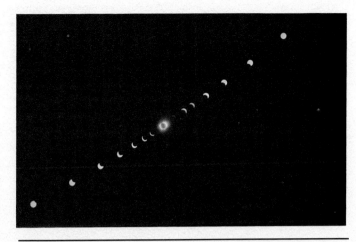

Above: A time-lapse sequence of the total eclipse of the Sun on July 10, 1972, seen from Prince Edward Island, Canada.

Left: A small telescope reveals the rugged, bright highlands of the Moon, interspersed with the smoother, dark lowlands known as maria. In this picture of the Moon, just past first quarter, the rounded Mare Serenitatis is at top right, with Mare Tranquillitatis below it.

Below: An amateur observer checks telescopes and cameras before photographing a total eclipse of the Sun.

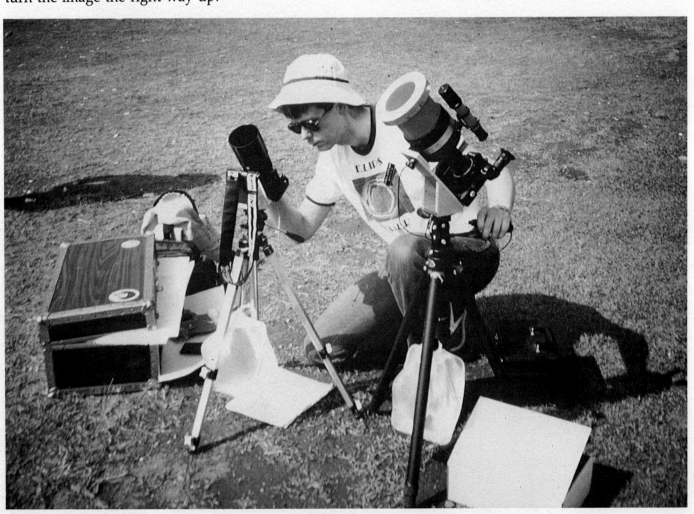

Above: As night falls, two amateur observers prepare for an observing session.

Left: The constellation of Orion. At top left is the red supergiant Betelgeuse, at lower right the blue giant Rigel. A line of three stars forms the belt of Orion, with the Orion Nebula below them. In the foreground is a 50-mm (2-inch) refractor, commonly used by beginners.

Telescope mountings

An important factor to look for in a telescope is its mounting. If this is not firm and stable, the telescope will be difficult to keep steady. There are two main types of telescope mounting. The simplest is known as an altazimuth mount, for it enables a telescope to swivel freely from side to side (in azimuth) and up and down (in altitude). This is the same design as used by coin-in-the-slot telescopes; it is cheap and simple, but the telescope has to be continually adjusted in two axes to keep the object in the field of view as the Earth turns.

Better suited for astronomy is the equatorial mount, in which one axis, called the polar axis, is parallel to the Earth's axis. Hence, as the Earth turns, the telescope needs only to be turned around this axis to keep it pointed at the object in view. Sophisticated telescopes have a motor that turns the polar axis at the same speed as the Earth rotates, leaving hands free for making notes and drawings.

Photographing the sky

You can use an ordinary camera to photograph the sky. The simplest type of astronomical photograph is a short time exposure with a fixed camera, which will show the brightest stars and planets. Fast film is best, though whether it is colour or black and white makes no difference. Fix the camera firmly somewhere so that it will not move during the exposure. Open the camera lens to its widest, set the shutter for a time exposure – using a cable release that can be locked to keep the shutter open – and focus on infinity. Then take an exposure lasting 10 seconds or so.

During longer exposures the stars will start to form short trails, caused by the rotation of the Earth. You can use this to take remarkable pictures which demonstrate the Earth's rotation. Point the camera towards the pole star, Polaris, and leave the shutter open for five to ten minutes. The resulting pictures will show star trails curving around the pole. If you leave the shutter open for too long, glare from streetlights will build up on the film and fog the picture. However if you are in dark country skies, away from streetlights, much longer exposures will be possible, showing longer star trails. For a change of scale, you can try using a telephoto lens. If you want to capture meteors (shooting stars), change to a wide-angle lens that covers a much greater area of sky.

A telescope can be used as a telephoto lens on the sky. For best results, remove the camera lens and fix the camera with an adaptor onto the end of the telescope. You can take high-magnification photographs of the Moon and planets through the telescope eyepiece, or remove the eyepiece for wider-angle views. However, for long exposures the telescope must be guided carefully or the photograph will be blurred. This requires accurate motor drives.

Right: The Sun's glowing corona becomes visible at a total eclipse, when the Moon blots out the Sun's surface. This eclipse, on February 16, 1980, was photographed from the Taita Hills in Kenya.

Below: Californian amateur Al Lilge opens up his observatory, housing a 32-cm (12.5-inch) reflector.

EXPLORING THE SOLAR SYSTEM

Our Sun, an average star, is surrounded by a family of nine planets and a host of associated debris that makes up the Solar System. The Sun and its family are believed to have been born from the same cloud of gas and dust in space around 4,600 million years ago. According to modern views, the formation of planets is a natural by-product of the birth of stars, so that there may be many other planetary systems in space – some of them, perhaps, harboring life.

The planets grew, it is thought, from a ring of leftover material around the primitive Sun. In this cloud, atoms of rock and metal were stuck together by collisions until they were big enough to attract one another by their own gravity. Thus, lumps tens or hundreds of kilometers across were produced, which merged into objects many thousands of kilometers in diameter – the planets. Remnants from this sweeping-up process are the chunks of rock and metal known as meteorites which occasionally collide with the Earth.

Meanwhile, the glowing fires of the young Sun were pushing the light gases out of the Solar System. Nearest to the Sun, where it was hottest, only the heaviest elements could remain. There,

dense planets of rock and metal were formed – Mercury, Venus, Earth, and Mars. Farther from the Sun, where it was colder, formed the giant gaseous planets of the Solar System – Jupiter, Saturn, Uranus, and Neptune. At the outermost reaches of the Solar System, frozen lumps of rock and ice were left – the comets. All remaining gases were blown away by the Sun's glare, dispersing into space. The Sun contains approximately 99.9 per cent of all the mass now remaining in the Solar System.

Heat from the decay of radioactive atoms warmed the interiors of the planets, so that light rocks rose to the top, forming a crust. Sufficient heat remains inside the Earth to cause occasional volcanic eruptions. Gases exhaled from the planets' interiors produced atmospheres, and much of the gas from volcanoes on Earth is water vapor, which condensed to form the seas. The smallest bodies, such as Mercury, did not have sufficient gravity to retain these gases, and so they remain airless and waterless.

Above: The Earth's atmosphere has been built up from gases erupted by volcanoes. The process continues today, albeit in diminished form, as in this eruption of Mount St Helens, USA, in 1980.

Left: Sunset on Mars, photographed by Viking 1 in 1976.

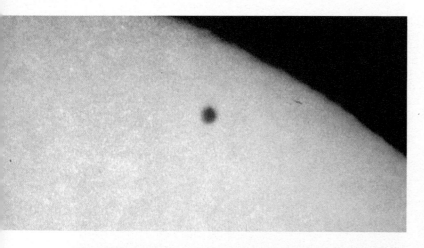

Above: The planet Mercury appears as a tiny black dot crossing the face of the Sun. Passages of Mercury in front of the Sun are seen from Earth every 10 years.

Mercury

Let us begin a tour of the Solar System at this tiny planet, the closest to the Sun. Mercury is the second-smallest planet in the Solar System (remote Pluto is smallest of all). Mercury's diameter of 4,880 km (3,030 miles) is only 50 per cent larger than our own Moon. For its size, Mercury is very heavy, and is believed to have a core of iron that extends four-fifths of the way to its surface. Fleet-footed Mercury orbits the Sun in a mere 88 days at an average distance of 57.9 million km (36 million miles). From Mercury, the Sun appears two-and-a-half times as large in the sky as it does from Earth. The Sun's heat is so great that temperatures on the day side of Mercury rise to 350°C, hot enough to melt tin and lead. On the dark side, the temperature of Mercury drops to −170°C.

Observing Mercury

Mercury is so close to the Sun that it is difficult to see with the naked eye except under the most favorable circumstances. Even in powerful telescopes it appears as little more than a tiny ball, and for a long time astronomers knew very little about it. For instance, they assumed that Mercury spun on its axis in the same time as it took to complete one orbit, so that it kept one face permanently turned towards the Sun, as the Moon does towards the Earth. However, in 1965 astronomers were surprised by results from the new technique of interplanetary radar, in which a radio beam is bounced off the surface of a planet; the radar reflection reveals both surface detail and the planet's speed of rotation. This revealed that Mercury rotates once every 59 days; in other words, it spins one-and-a-half times on its axis every time it orbits the Sun. Therefore, for the Sun to go once around the sky as seen from the planet's surface – say, from one noon to the next – Mercury must complete two orbits, spinning three times on its axis, which takes a total of 176 Earth days.

In 1974, astronomers got their first good look at the planet when the US Mariner 10 space probe sent back photographs which showed that Mercury looks strikingly similar to the Moon. Its rocky surface is pockmarked with giant craters, apparently caused by a bombardment of debris shortly after the formation of the Solar System. The largest scar on Mercury is the Caloris Basin, 1,400 km (800 miles) in diameter, which has apparently been flooded by molten lava, like the lowland plains of the Moon. The most recent craters are small, bright pits, often surrounded by bright rays of ejected material. This barren, rugged surface can harbor no form of life. Humans will not travel to this forbidding planet for a long time to come.

Venus, the Hell planet

This planet, second in line from the Sun, was for a long time even more mysterious than Mercury. Venus orbits the Sun at a distance of 108.2 million km (67.2 million miles), and at its closest can come to within 38 million km (24 million miles) of the Earth, closer than any other planet. Its surface is shrouded from view by an unbroken blanket of white clouds which reflect most of the sunlight that strikes them, so through any telescope the planet appears as little more than a dazzlingly bright ball that goes through a series of phases as it orbits the Sun every 225 days. At its brightest, Venus is familiar as the morning or evening 'star', far outshining all other stars and planets.

Its diameter of 12,100 km (7,520 miles), similar to that of the Earth, earned it the nickname 'Earth's twin', but, alas, it has turned out to be very different. During the 1960s, radio astronomers detected radiation from the planet which indicated that its surface temperature was very high. Since then, space probes have parachuted to the surface, confirming that beneath its clouds Venus resembles Hell. Its atmosphere is made of unbreathable carbon dioxide gas, which at the surface presses down with 90 times the force of the Earth's atmosphere, equivalent to pressures a kilometer deep in the ocean. In addition, the dense atmosphere traps the Sun's heat like a greenhouse, building up a furnace-like surface temperature of 475°C. The bright clouds, which lie at a height of about 65 km (40 miles), are made not of

Top right: At first sight, the cratered surface of Mercury, as photographed by the Mariner 10 probe in 1974, looks like that of our own moon.

Right: Clouds swirl around Venus, permanently blanketing its surface from outside view. This picture was taken by the Pioneer-Venus orbiter probe in 1979.

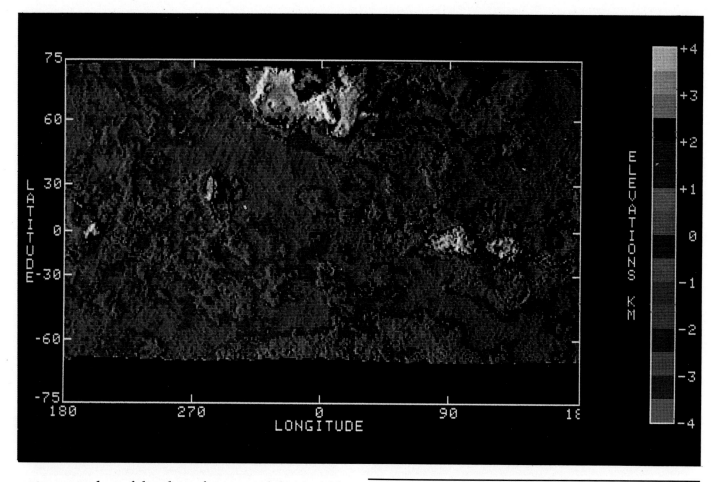

Above: Radar is used to build up maps of Venus, such as this one made by the Pioneer-Venus orbiter, showing two main continental areas.

water vapor but of droplets of strong sulphuric acid. Beneath them, a rain of sulphuric acid falls. Certainly, nothing living is likely to be found on or around Venus – nor is it a welcoming place for astronauts.

During the 1960s, interplanetary radar, which is not blocked by the clouds, revealed a remarkable fact: the planet rotates in the wrong direction, from east to west instead of from west to east as do the Earth and other planets. The time taken for one rotation, 243 days, is longer than Venus takes to orbit the Sun. How the rotation of Venus was thrown into slow reverse gear remains a puzzle. While the planet rotates slowly, the clouds of Venus move much more quickly. They swirl around the planet from east to west every 4 days, driven by high-altitude winds of 360 km/hour (225 miles/hour), faster than in a hurricane on Earth. In the dense lower atmosphere, though, winds are sluggish.

Probing the surface
Radar has enabled astronomers to map the surface of the planet, both from Earth and from the Pioneer space probe that went into orbit around Venus in December 1978. Being so hot, of course, Venus has no seas, but there are two main continental areas, with additional smaller highlands that may be volcanic mountains. Enormous bolts of lightning flash down from the sulphuric acid clouds to the mountain tops. One continental area, called Aphrodite, is cut by a rift valley over 1,000 km (600 miles) long. The rest of Venus is covered with rolling plains, dotted with craters.

On-the-spot views from the surface of Venus were first taken in 1975 by the Soviet lander craft Venus 9 and Venus 10. Their photographs showed that Venus is not as gloomy under its clouds as anticipated – it is as bright at the surface as on an overcast day on Earth. One lander came down at the foot of a rock slide, while the other found itself on a plateau with stony outcrops. Color photographs of the surface, taken in March 1982 by the Soviet Venus 13 and Venus 14 probes, show that under its clouds Venus is bathed in a sulphurous orange glow. The rocks on Venus are similar in composition to volcanic basalt, like those on the Moon's lowland plains. Internally, Venus is believed to be like the Earth – a heavy core surrounded by a mantle of lighter rock and topped with a thin crust.

The Moon
Every 27.3 days we are orbited by this rocky body, 3,476 km (2,160 miles) in diameter, whose average distance of 384,400 km (239,000 miles) makes it by

far our closest permanent natural neighbor. The Moon also takes 27.3 days to rotate once on its axis, so that it keeps the same face permanently turned towards the Earth. This has come about because the gravitational pull of the Earth has decelerated the Moon's spin into a 'captured' rotation. In turn, the Moon's gravity affects the Earth, producing tides in the oceans and slowing the Earth's rotation by 0.001 seconds per century. Although this is minimal, the effect builds up over long periods; the daily growth bands of fossil corals reveal that 380 million years ago there were 22 hours per day, and 400 days in a year.

Each month, the Moon goes through its familiar cycle of phases from new, through crescent, half, gibbous, to full, and back again to new. These phases are caused as the Moon orbits the Earth and we see different amounts of its illuminated portion. Since the Moon's path is inclined at 5° to the Earth's orbit around the Sun, only occasionally do the three bodies come in line to cause an eclipse. At least twice a year the Moon passes in front of the Sun to cause a solar eclipse, and about as often the Moon passes into the Earth's shadow, producing a lunar eclipse.

Even a casual glance at the Moon with the naked eye shows the dark lowland plains which form the so-called 'man in the Moon' pattern. Galileo termed

Below: Full Moon. The light areas are the cratered highlands, the dark areas are the lowlands. Bright rays splash outwards from some craters.

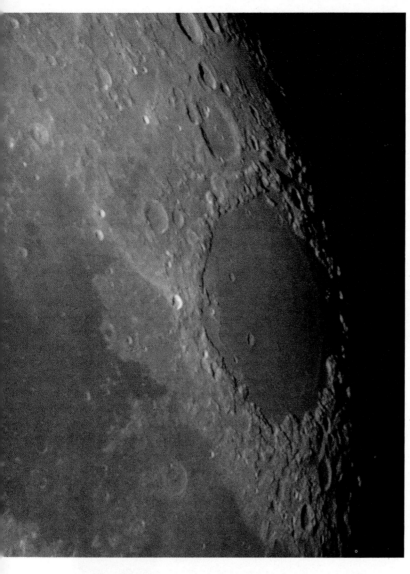

Left: Close-up of Mare Crisium, a nearly circular plain about 500 km (300 miles) in diameter.

these dark areas *maria* – Latin for seas – because he thought they really did contain water. We now know that the Moon is airless and waterless, but the name has stuck. Binoculars or a small telescope will disclose countless craters pitting the Moon's surface, particularly in the bright lunar highlands. After centuries of controversy, it is now accepted that the craters were formed by meteorite impacts, rather than by volcanoes as was once thought. There are only a few minor examples of volcanic activity on the Moon.

Discoveries of the Moon missions

Astronomers have mapped the visible hemisphere of the Moon in detail, naming major formations after famous scientists and, in some cases, historical or mythical characters. The far side of the Moon remained unseen and mysterious until 1959, when it was first photographed by the Soviet probe Luna 3. Following that, a series of unmanned American lunar orbiting craft photographed the entire Moon, front and back, in greater detail than ever before.

Surprisingly, the Moon's averted hemisphere is quite different in appearance from the side with which we are familiar: instead of large mare areas, it is mostly bright uplands, saturated with craters of all sizes. Evidently, the crust on the Earth-facing hemisphere is thinner, so that the dark lavas could burst through to form the maria.

Between 1969 and 1972, six Apollo crews landed on the Moon and brought back 380 kg (840 lbs) of rocks. Apollo experiments on the surface and from orbit expanded the knowledge gleaned by the first unmanned pathfinder probes, enabling astronomers to piece together a clear picture of the Moon's history. Overall, the Moon is less dense than the Earth; while the Moon rocks are broadly similar to volcanic basalt on Earth, there are subtle differences in composition which set them apart. Old ideas that the Moon split from the Earth have now been abandoned in favour of a separate formation for the Moon, but close enough for it to be tied to the Earth by gravity.

How the Moon evolved

Although the Moon is now cold inside, shortly after its birth it must have been much warmer. The lightest rocks rose to the top, forming a crust, while denser materials settled to the center, forming a core. For several hundred million years the lunar crust was subjected to a relentless bombardment from debris left after the solar system's formation. Impact piled upon impact, pulverizing the crust, until the storm abated. Between 4,000 million and 3,000 million years ago, deep basins were dug by the largest impacts into which molten lava seeped from below to produce the flat, dark maria. During this period on Earth, continents were being built and the first life appeared. Since then, the face of the Moon has been disturbed only by the occasional encounter with a stray meteorite, and a continual peppering by micrometeorites like cosmic hailstones that churn the rocks into fine soil. However, sometime next century it will be disturbed when humans return to the Moon to set up bases and perhaps to mine its crust for the iron, titanium, magnesium, and other minerals in which it is so rich.

Right: Ten-day Moon.

Far right: Waning, 19-day old Moon. The bright, rayed craters are Tycho (bottom), Copernicus (just left of center), Kepler (left of Copernicus) and Aristarchus (the very bright spot towards top left).

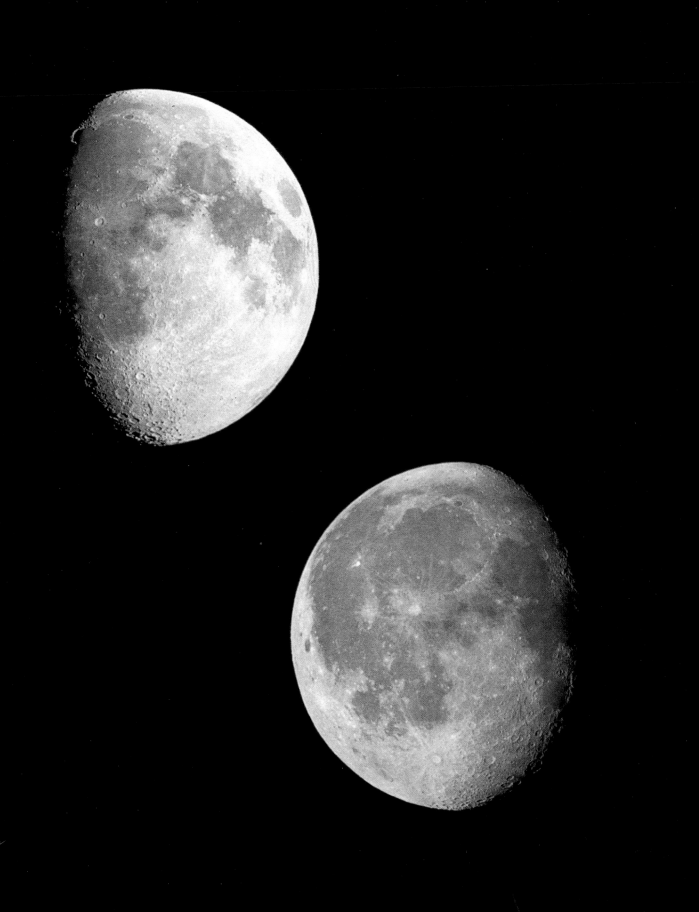

Mars, the red planet

Fourth in line from the Sun, Mars was once thought likely to harbor other life. Popularly known as the red planet because of its distinctive color, actually caused by large amounts of reddish-brown iron oxide in its crust, Mars orbits the Sun every 687 days at an average distance of 227.9 million km (141.6 million miles). Although its diameter of 6,794 km (4,220 miles) is only about half that of our home planet, Mars has many superficial similarities with the Earth. For instance, it rotates on its axis once every 24.6 hours, giving it a day scarcely longer than our own. It has a thin atmosphere in which clouds form, and two glistening white polar caps that melt during the summer and re-form each winter.

Martian discoveries

A telescope of at least 150 mm (6 inches) aperture is needed to observe Mars. Several dark markings appear greenish in hue by contrast with the bright red deserts surrounding them, and which astronomers once thought might be expanses of vegetation. During the close approach between Earth and Mars in 1877, the Italian astronomer Giovanni Schiaparelli (1835–1910) detected a number of straight markings which he termed *canali*, meaning channels. The word was mis-translated as canals, implying that they were of artificial construction, which Schiaparelli did not believe. Percival Lowell (1855–1916), a wealthy American astronomer who set up his own observatory at Flagstaff, Arizona, became firmly convinced that the canals were artificial waterways dug by Martians to bring water from the polar caps to irrigate their crops at the equator. But other astronomers failed to see the network of fine lines drawn by Lowell. Instead, under the best conditions, the so-called canals seemed to break up into disjointed dots and splodges. The canal controversy was finally laid to rest by the close-up views afforded by space probes, which confirmed that the canals do not exist. Lowell and his followers had been deceived by tricks of the eye.

Right: Morning mist in the canyon area known as Noctis Labyrinthus on Mars.

Left: Effects of weathering and lighting give this mesa the appearance of a face. It is about 1.5 km (1 mile) across and lies in the Elysium area of Mars.

Below: Valles Marineris, part of the extensive 'grand canyon' of Mars produced by a combination of surface faulting and wind erosion. Its total length is 4,000 km (2,500 miles).

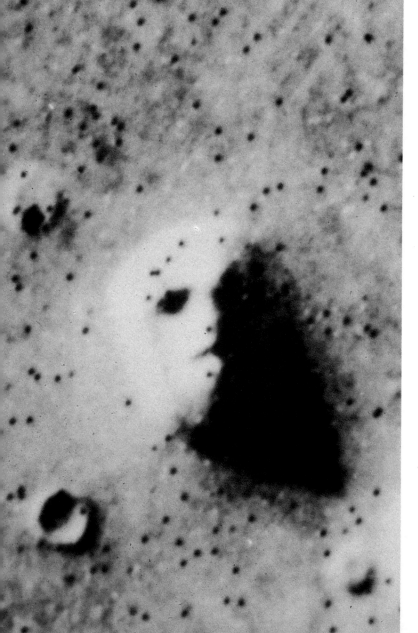

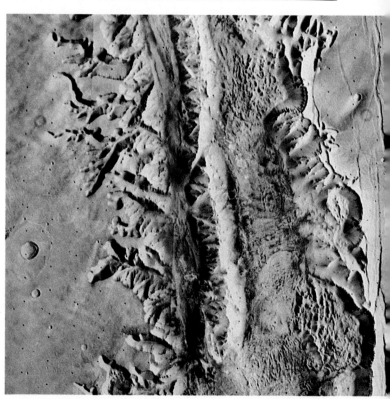

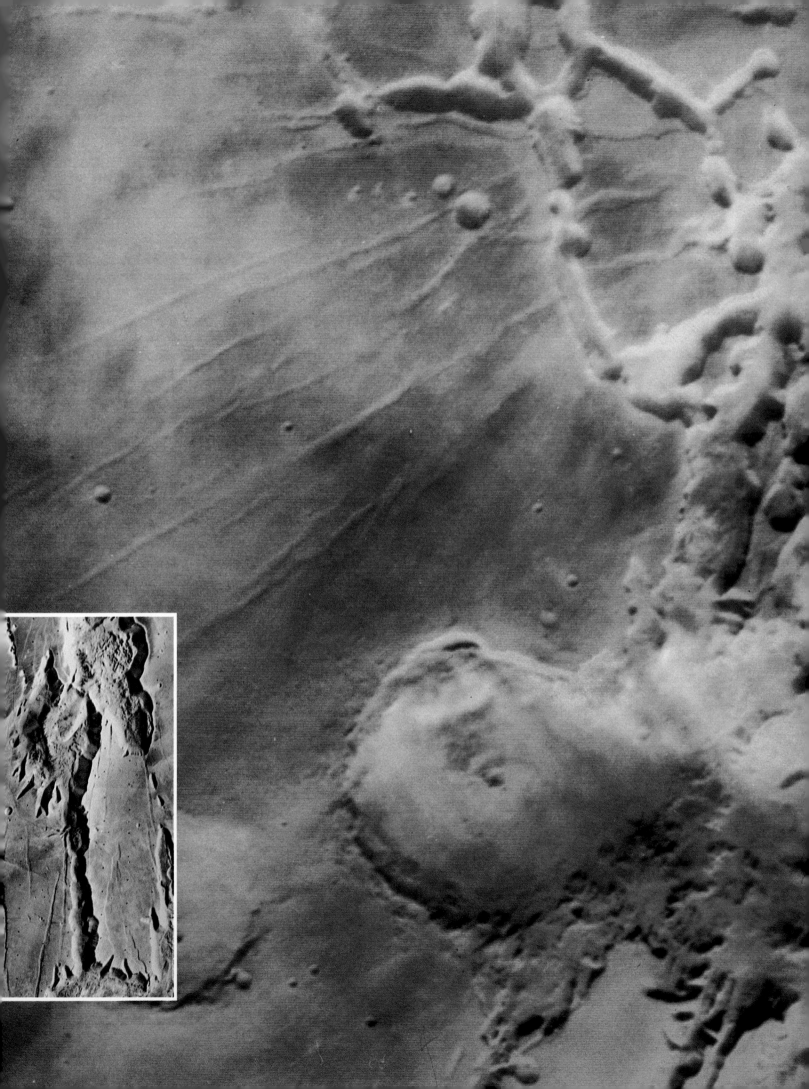

The Mariner and Viking missions

What the space probes did find were craters like those on the Moon, relics of the sweeping-up process that followed the formation of the Solar System. The probes also found that the atmosphere of Mars was composed of carbon dioxide and was far thinner than expected – as thin as the Earth's atmosphere at a height of 32 km (20 miles) – meaning that Mars must be very cold. A complete photographic survey of Mars from orbit by the American space probe Mariner 9 in 1971–72 revealed a gigantic volcano known as Olympus Mons, 600 km (375 miles) wide and 24 km (15 miles) high, making it the largest known volcano in the Solar System, bigger even than the volcanic Hawaiian islands on Earth. Also revealed was a canyon similar to the African rift valley, 4,000 km (2,500 miles) long and 120 km (75 miles) wide. The discovery of what appeared to be dried-up river channels on the surface increased optimism that the planet's climate may in the past have been sufficiently warm and damp for some primitive life-forms to have arisen. Two Viking spacecraft were dispatched to find out.

When the Viking landers set down on Mars in the summer of 1976, their cameras revealed a rust-red landscape with no signs of life. Fine dust particles suspended in the atmosphere turned the Martian

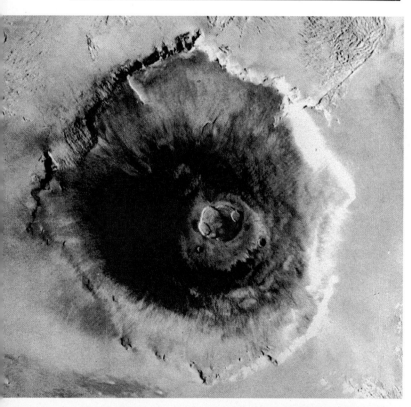

Below: Olympus Mons is a huge, extinct volcano on Mars, 600 km (375 miles) across. It has a complex crater formed after eruptions; lava flows are spread out on the surrounding plains.

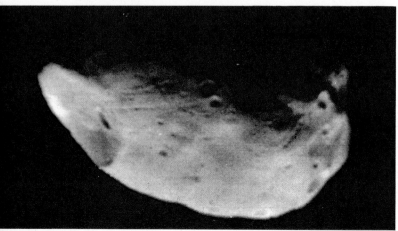

Bottom: Phobos, the larger of the two potato-shaped moons of Mars, is 28 km (17 miles) across at its widest. It may be a captured asteroid.

sky pinkish-red. The Vikings recorded the true harshness of the Martian environment: even on a sunny summer's afternoon the air temperature rises to only -30°C; at night, it falls to -86°C. The polar caps are made of a mixture of water ice and frozen carbon dioxide (dry ice), and there is probably a permanently frozen layer (permafrost) under the Martian surface.

Each Viking lander carried a scoop to sample the Martian surface and a miniature biology laboratory to incubate the soil in search of Martian microorganisms. Alas, although the soil produced many interesting chemical reactions, the Vikings found no trace of even the tiniest bugs on Mars. The red planet seems to be sterile after all. Nonetheless, one day there will be life on Mars – when the first humans visit it. Such an expedition will not take place until well into next century.

Mars has two tiny moons, Phobos and Deimos. They were discovered by the American astronomer Asaph Hall (1829–1907) during the same close approach of Mars in 1877 at which the canals were reported. Phobos, the larger of the two, is so close to Mars, a mere 6,000 km (3,700 miles) above the surface, that it crosses the Martian sky twice a day. Deimos is three times more distant. Space probe photographs show that Phobos and Deimos are irregularly shaped lumps of rock, probably stray bodies captured early in the history of the Solar System. Their respective diameters are approximately 22 km (14 miles) and 12 km (8 miles).

Below: The surface of Mars is a rock-strewn red desert. The sky is pink, caused by red dust suspended in the atmosphere. This picture was taken by the Viking 2 lander in 1976.

Mighty Jupiter
This planet, the giant of the Solar System, weighs two-and-a-half times as much as all the other planets put together. Jupiter is 142,800 km (88,700 miles) in diameter, 11 times that of Earth. It lies in the cold outer reaches of the Solar System, 778.3 million km (483.6 million miles) from the Sun, and takes 11.9 years to complete one orbit. Jupiter is different in nature from the small, rocky bodies that make up the Solar System – it is a gas giant, composed mostly of hydrogen and helium, the same gases as the Sun. In fact, had Jupiter been about 100 times bigger it would have become a small star. Since Jupiter's gravity is so great – the escape velocity is 5.3 times that of the Earth – it has been able to retain even the lightest gases, and so has changed little since its formation.

Even through small telescopes with apertures of

Above: Colorful cloud belts of the giant planet Jupiter as photographed by Voyager 1 in January 1979 from a distance of 35 million km (22 million miles).

Right: Close-up from Voyager 2 of the Great Red Spot, a storm system three times the width of the Earth. Twisted ribbons of cloud swirl around it; below are two smaller white spots.

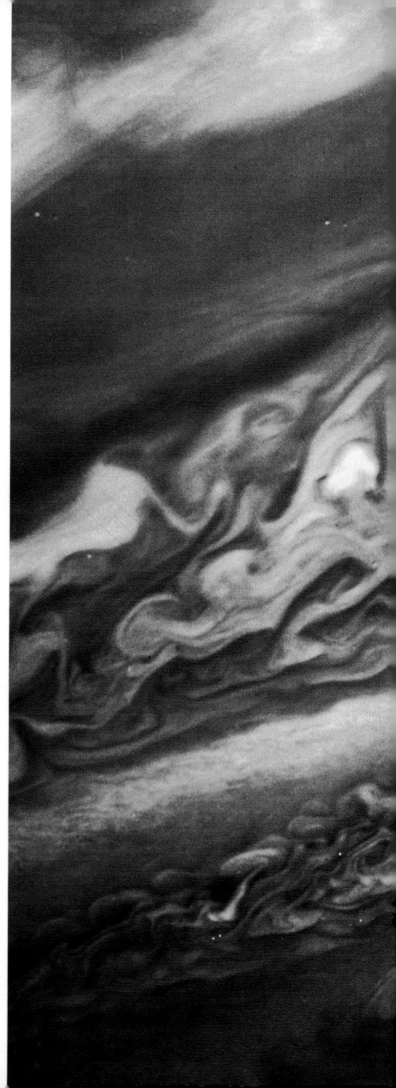

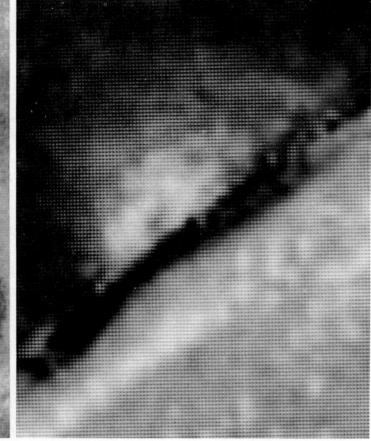

Above: Jupiter's moon Io is the most volcanically active body in the Solar System. Its volcanoes erupt sulphur, turning Io's surface yellow and orange. **Below:** A volcano erupting at the edge of Io is seen by the space probe Voyager 1 in March 1979.

no more than 75 mm (3 inches), Jupiter presents an abundance of detail. Multi-colored cloud belts stretch around the planet, drawn out into parallel bands by Jupiter's high-speed rotation of just under 10 hours, the fastest of any planet. Traces of methane, ammonia, and other chemicals produce the colors from yellow, through red and brown, to purple and blue, seen among the continually changing spots and swirls in the Jovian clouds. The only permanent feature in Jupiter's clouds is the Great Red Spot, an enormous oval the width of three Earths, photographed in astounding close-up by the Voyager 1 and 2 space probes that sped past the planet in 1979.

What the Voyagers discovered
The Great Red Spot seems to be a scaled-up version of the anvil-shaped clouds seen over thunderstorms on Earth, caused by a permanent updraught of hot gas from the planet's interior; the spot is actually the highest cloud on Jupiter. Red phosphorus welling up from deep in the Jovian atmosphere is thought to give the spot its characteristic color. One surprising fact confirmed by space probes is that Jupiter releases two-and-a-half times as much heat as it receives from the Sun. This internal heat, thought to be left over from its formation, stirs the continually changing cloud patterns on Jupiter.

There is probably no such thing as a solid Jupiter beneath the clouds. At the cloud tops the temperature is -150°C, but lower levels of the atmosphere may be warm enough to support some kind of airborne life. About 1,000 km (600 miles) below the visible cloud tops, pressures become so great that hydrogen is compressed into a liquid. Towards the planet's center the hydrogen is squeezed even further so that it becomes electrically conducting, like a metal. Possibly Jupiter has a rocky core about the size of Mars. Convection currents in the electrically conducting hydrogen produce a strong magnetic field around Jupiter which traps atomic particles from the Sun, producing radiation belts with several hundred times the radiation dose necessary to kill a man.

The Jovian family
Astronauts will never be able to land on Jupiter, but they might visit several of its numerous satellites, which form a kind of mini solar system. Ganymede, the largest satellite in the Solar System, has a diameter of 5,275 km (3,278 miles), larger than the planet Mercury, and Callisto is not much smaller. Io and Europa, the other two major satellites, are similar in size to our own Moon. All four, discovered by Galileo, can be seen in binoculars. Io is the most volcanically active body in the Solar System because it is continually squeezed by the opposing gravitational forces of Jupiter and the

other nearby satellites, which releases heat that keeps its interior molten. Volcanoes are continually erupting on its surface, spewing out flows of sulphur that give Io a mottled orange colour. Europa, by contrast, is encrusted with ice, giving it a smooth white surface like a billiard ball. Under its ice, Europa may be warm enough to possess an ocean of liquid water, possibly harboring some form of life.

Saturn, the ringed planet
This, the most beautiful planet of all, is the outermost planet visible with the naked eye, 1,429 million km (888 million miles) from the Sun and taking 29.5 years to complete one orbit. Saturn, another gas giant composed mostly of hydrogen and helium, is noted for the rings which encircle its equator. Made of rocky particles on average about the size of bricks, they are coated with frozen water and orbit Saturn like a swarm of moonlets. From rim to rim the rings span 275,000 km (170,000 miles), yet they are a mere 100 meters thick from top to bottom. Therefore, in relation to their diameter, they are thinner than a sheet of tissue paper. Probably the rings represent the building blocks of a potential moon that never formed, or they could have been produced when a former moon of Saturn strayed too close to the planet and was broken up by its strong gravitational force.

Saturn is the second-largest planet in the Solar System, 120,000 km (74,500 miles) in diameter. Its rotation period, 10.25 hours, is second only to that of Jupiter. Saturn's cloud belts are less distinct and less turbulent than those of Jupiter, and there is no equivalent of the Red Spot. Although Saturn is similar in composition to Jupiter, the gases of which it is made are not compressed as tightly, so its overall density is less than that of water; given a big enough ocean, it would float.

Below: Space probes to the planets are controlled from this room at the Jet Propulsion Laboratory, Pasadena, California, seen here during Voyager 1's approach to Saturn in 1980.

Left: Voyager 2 looked back on Saturn and its rings after it had passed the planet in 1982.

Right: Under the close-up inspection of the Voyager cameras, the rings of Saturn break up into thousands of thin strands. For scale, the Earth is superimposed.

Below: The beautiful ringed planet Saturn photographed by Voyager 1 in 1980 from a distance of 51 million km (32 million miles).

Observing Saturn

Small telescopes will show the rings of Saturn. In larger telescopes the bright outer part of the ring is seen to be separated from the dimmer inner part by a 3,000-km (2,000-mile) gap, named Cassini's division after the French astronomer Jean Dominique Cassini (1625–1712), who detected it in 1675. On the inner side of these rings is the transparent crepe ring. Saturn's rings are actually far more complex than they appear from Earth, as the two Voyager space probes discovered when they flew past the planet in 1980 and 1981. In close-up, the rings are seen to consist of thousands of narrow ringlets, like the ridges and grooves of a long-playing record.

Saturn has about 20 known moons, more than any other planet. Among them is Titan, the only moon to possess a substantial atmosphere, consisting mostly of nitrogen with some methane, topped by orange clouds. Titan's diameter of 5,120 km (3,180 miles) ranks it second in size to Jupiter's largest moon, Ganymede. Under its clouds, Titan may be covered with a sea of liquid methane, which will only be ascertained by a future space probe. Another moon of Saturn, Mimas, sports an amazing

crater 135 km (84 miles) in diameter, fully one-third the size of the moon itself. Mimas must have been virtually shattered by the force of the impact that created the crater.

Uranus

In 1781, Sir William Herschel doubled the known size of the Solar System. While scanning the skies, he accidentally discovered the planet Uranus, 2,875 million km (1,786 million miles) from the Sun, twice as far as Saturn and only just visible to the naked eye. Uranus is remarkable in that it seems to have fallen on its side; its axis of rotation lies almost exactly in the plane of its orbit, so that each pole is subjected to 42 years of perpetual sunlight, followed by 42 years of perpetual darkness, as the planet completes its 84-year orbit around the Sun. No one knows the reason for this exaggerated tilt, but it must have come about during the planet's formation, perhaps as the result of a collision.

Uranus appears greenish because the hydrogen and helium of its atmosphere are mixed with considerable amounts of methane, which absorbs red light. Beneath the featureless green clouds, over half the 52,000 km (32,300 miles) diameter of Uranus is made up of a core of rock and ice, mostly frozen water and ammonia. It spins on its axis once about every 16 hours.

Uranus has more than five moons, which orbit its equator so that they, too, are highly inclined to the plane of the planet's orbit. Between 16,000 km and 25,000 km (10,000 to 15,500 miles) from the cloud tops lie nine thin rings of rocky debris, discovered in 1977 when they blocked out light from a star as Uranus passed in front of it. They are probably the remains of a shattered former moon, composed of very dark material and so faint that they are scarcely visible directly, which is why they were not seen before. The eight inner rings are no more than 10 km (6 miles) wide, while the outermost ring has a width of 100 km (60 miles).

Neptune

After Herschel's discovery of Uranus, astronomers found that the new planet was not keeping to its predicted path. Something seemed to be pulling it out of line – perhaps another as yet undiscovered planet. Mathematicians started to calculate its possible position; in England, John Couch Adams (1819–1892), and in France Urbain Leverrier (1811–1877), reached almost identical conclusions, and on September 23, 1846, the German astronomer Johann Gottfried Galle (1812–1910) found the new planet close to the predicted position.

Neptune orbits the Sun every 165 years at an average distance of 4,500 million km (2,800 million miles). Like Uranus it appears as a green, featureless disk in a telescope, and in fact the two planets are thought to be very similar in composition and structure. Neptune is slightly the smaller of the two, being 48,500 km (30,100 miles) in diameter. Its rotation period, still not known with certainty, is about 18 hours, and it is thought to have a set of faint rings similar to those of Uranus.

Neptune has two satellites, both extraordinary: Triton, 3,200 km (2,000 miles) in diameter, orbits Neptune from east to west, the reverse of the normal west-to-east traffic pattern in the Solar System. Triton's orbit seems to be getting smaller so that eventually it will come so close to Neptune that it will break up, forming more rings. Nereid, the more distant of the two moons, moves in the standard west-to-east direction, but its orbit is the most elongated of any moon, ranging between 1.4 million and 9.7 million km (0.9 to 6 million miles) from the planet.

Pluto

After the discovery of Neptune it was natural that astronomers should go hunting for other possible planets. Most dedicated of them was Percival Lowell (of Martian canal fame), who instituted a photographic search for a trans-Neptunian planet at his observatory in Arizona. Not until long after Lowell's death did the search produce results, when Clyde Tombaugh (b.1906), an assistant at the observatory, discovered Pluto on February 18, 1930, a planet so slow-moving that it takes 248 years to complete one orbit of the Sun, and so will not return to its discovery position until 2178.

Pluto's diameter is now known to be no more than 3,000 km (1,900 miles), less than our own Moon, and hence the smallest planet of all. It rotates slowly on its axis, once every 6 days 9 hours. Part of its surface is covered with frozen methane, at a temperature of -230°C, and seems to have more in common with Neptune's largest moon, Triton, than it does with any of the planets. From Pluto, the Sun appears as little more than a brilliant star.

In 1978 astronomers discovered that Pluto has a

Above: Artist's impression of the planet Pluto and its moon Charon, in the dark outer reaches of the Solar System.

Left: Artist's impression of Neptune as it might appear from its largest moon, Triton.

Right: A computer-enhanced photograph of a disk of gas and dust around the star Beta Pictoris, believed to be forming into planets.

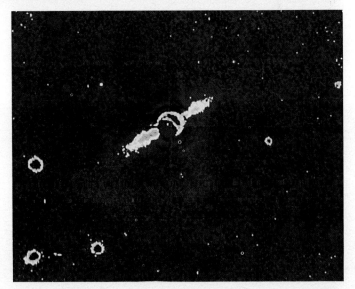

moon of its own, which they named Charon; it has a diameter of about 1,000 km (600 miles), making it the largest satellite relative to its parent planet in the Solar System. Charon orbits Pluto in the same time that the planet takes to spin on its axis, so that it remains fixed over one point on Pluto's equator like a geostationary satellite above the Earth.

Pluto's average distance from the Sun is 5,900 million km (3,600 million miles), but its orbit is eccentric – at its closest it can come within the orbit of Neptune. This is happening now – between January 1979 and March 1999 – during which time Neptune is the outermost planet of the Solar System. Pluto is the only planet to cross another's orbit, leading some astronomers to suppose that Pluto is an escaped satellite of Neptune. However, it now seems more likely that Pluto has always been a separate body, a potential third moon of Neptune that was never captured. Are there other planets beyond Pluto? Probably not. Any other planets would have to be extremely small, or extremely distant, or both, to have escaped detection by now.

Other planetary systems

Since our own Sun has planets, it seems likely that many other stars have planets also, and astronomers are now discovering signs of them. In 1983 the Infra Red Astronomy Satellite, IRAS, detected disks of cool dust around 40 nearby stars, and soon afterwards astronomers were able to photograph the disk around one of these stars, Beta Pictoris. Our own planetary system is thought to have arisen from such a disk of dust and gas encircling the Sun, so what we are seeing around these other stars is probably the formation of planets of their own. Astronomers estimate that one in ten of all stars may be accompanied by planets, some of which may have given rise to life.

SKY WANDERERS

Beyond Mars lies a belt of rubble known as the asteroids, too faint to be seen without a telescope. Ceres, the largest, 1,000 km (600 miles) in diameter, was discovered in 1801 by the Italian astronomer Giuseppe Piazzi (1746–1826); now, thousands are known. An estimated total of 100,000 asteroids may be visible in the largest telescopes, although most of these will be no more than a few hundred meters across. Typical asteroids probably resemble Phobos and Deimos, the moons of Mars. Some with eccentric orbits wander outside the normal asteroid belt; Hermes, for example, came within 800,000 km (500,000 miles) of the Earth in 1937. But most asteroids orbit between Mars and Jupiter.

If all the asteroids were gathered together they would make a body only one thirtieth the mass of the Moon. They do not, therefore, represent the remains of a shattered former planet as was once suggested. They are probably examples of the primitive bodies of the Solar System from which the planets formed, but were themselves prevented from growing into anything larger by the overshadowing gravitational pull of mighty Jupiter.

Below: A small space probe heaves to among the asteroids, a possible source of raw materials in centuries to come.

Comets

Beyond Pluto there is probably nothing more than a belt of millions upon millions of comets, the ghostly wanderers of the Solar System which loop around the Sun on highly extended orbits lasting up to millions of years. The comet belt lies at a distance of about one light year from the Sun, roughly a quarter the way to the nearest star. The gravitational influence of passing stars nudges comets out of this cloud and towards the Sun where they can become trapped by the gravitational pulls of the planets into much shorter orbits, such as Comet Encke, which has the shortest known period, 3.3 years. Each year about 10 new comets are discovered, many of them by amateur astronomers who keep a special watch with telescopes and powerful binoculars. Anyone who discovers a comet can have it named after them.

When far from the Sun a comet resembles a dirty snowball – a bag of rocks and dust cemented into a ball a few kilometers across by frozen gas. As the comet plunges in towards the Sun it warms up and the gases evaporate to form a glowing head up to

Above: Comet Bennett appeared in 1970. This photograph was taken with a 3-minute exposure through a home-made 41-cm (16-inch) reflecting telescope.

Above: inset: Halley's Comet moves in an elliptical orbit, bringing it close to the Sun every 76 years. Its last appearance was in 1986; its next will not be until 2061.

100,000 km (60,000 miles) across, and a long, flowing tail which in the case of Halley's comet in 1910 stretched for 150 million km (93 million miles), as far as the Earth is from the Sun. Halley's comet, which orbits the Sun every 76 years, made its last return in 1985–86, when it was greeted by a flotilla of five space probes. Named after the English astronomer Edmond Halley (1656–1742), who calculated its orbit in 1705, it moves from between the orbits of Mercury and Venus out to beyond the orbit of Neptune. Records of its appearances go back to 240 BC. Despite their apparent grandeur, comets are so insubstantial that a million million of them would be required to outweigh the Earth.

Meteors and meteorites

Comets lose gas and dust each time they approach the Sun, eventually fading out. Dust particles from comets are constantly being swept up by the Earth; they burn up in the atmosphere by friction to form the darting streaks of light known as shooting stars or meteors. Meteors are the size of grains of sand, too small to reach the ground before burning up.

Several times a year the Earth encounters swarms of dust from comets, and a meteor shower results. The members of a particular meteor shower appear to be radiating from one part of the sky, after which the shower is named – for example, the Perseids appear to radiate from Perseus, the Geminids from Gemini, and so on. Amateur astronomers can make useful observations of meteor showers simply by counting the number of meteors seen per hour and estimating the brightness of each one. Two of the brightest meteor showers are the Perseids, which reach maximum activity around August 12 each year, and the Geminids, which peak around December 14. In both cases, several dozen meteors an hour can be seen by the naked eye in clear skies.

Occasionally, much larger chunks of rock and metal – meteorites – penetrate the atmosphere. These do not come from comets but are probably fragments of asteroids. At Hoba West in Namibia lies the heaviest known meteorite, a chunk of iron and nickel weighing 70 tonnes. If a meteorite hits the Earth at high enough speed it will blast out a crater like those on the Moon. A giant crater over 1 km across in the Arizona desert was formed about 25,000 years ago by the impact of an iron meteorite weighing a quarter of a million tonnes, which exploded with sufficient force to devastate a city. Some scientists have speculated that an impact with a comet or asteroid could have rendered the dinosaurs extinct 65 million years ago. Fortunately, encounters with such bodies are rare.

Right: Meteors flash out during the Leonid meteor shower of 1966, shown as vertical streaks. The curved lines are star trails, drawn out as the Earth rotated during the time exposure.

Right inset: Objects from space sometimes crash, blasting out scars on the Earth's crust. This circular lake 70 km (43 miles) in diameter at Manicouagan in Quebec, photographed from space, was caused by the impact of a large meteorite about 200 million years ago.

Top right inset: Meteor Crater in Arizona, created by the impact equal to that of a 10-megaton bomb, of an iron meteorite about 25,000 years ago. The crater is 1.2 km (4,200 ft) in diameter, 200 m (600 ft) deep.

FURNACES IN THE SKIES

Life on Earth would be impossible without the Sun, our prime energy source. The Sun's heat keeps temperatures on Earth warm enough for life, and its light is used as energy by plants to grow. When we burn fossil fuels – coal, oil, and natural gas – we are actually releasing the stored energy of sunlight.

The Sun is a glowing ball of gas of awesome size – its diameter of 1,392,000 km (865,000 miles) is equivalent to a row of 109 Earths. At our distance of 149,600,000 km (93 million miles) the Sun appears comfortably warm and bright. If we were closer to the Sun, at the distance of Mercury or Venus, conditions would be too hot for life, whereas farther away, at the orbit of Mars or beyond, sunlight would be too weak.

From our uniquely favorable position in orbit around the Sun we can observe its surface in some detail. One important warning to bear in mind is that you should *never* look directly at the Sun through any form of optical instrument; even

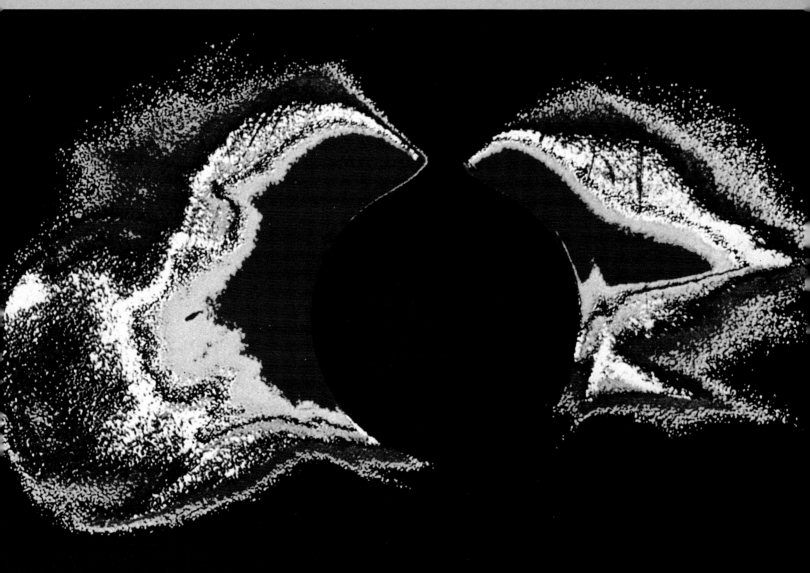

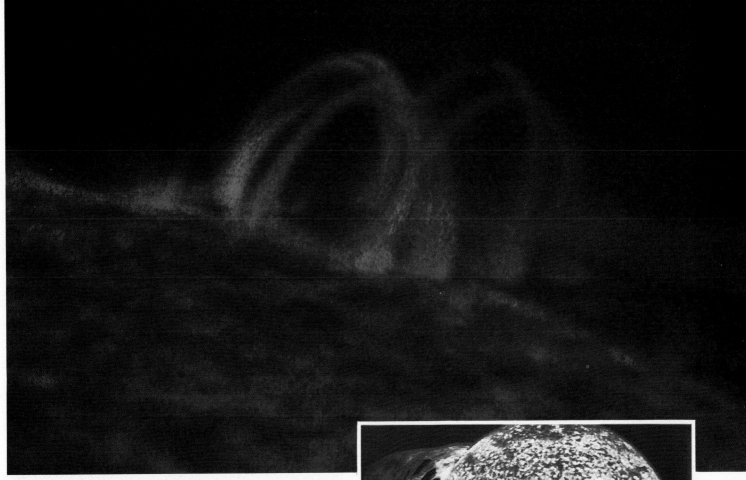

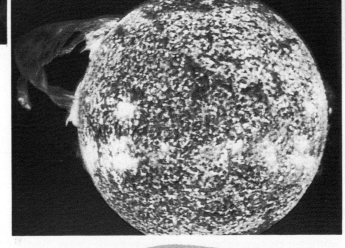

Above: Loops of gas follow magnetic fields above a sunspot.

Right: An immense prominence of glowing gas loops out 500,000 km (300,000 miles) from the surface of the Sun, photographed in ultraviolet light from Skylab in 1973.

Left: Sun's corona as seen in an image-processed photograph from Skylab.

Below: Sunspots are caused by strong magnetic fields bursting through the Sun's surface.

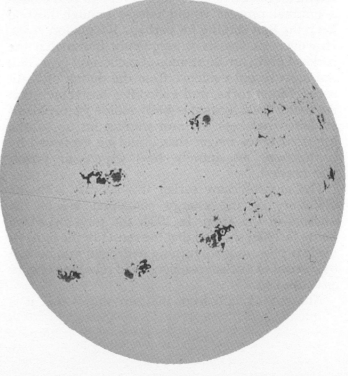

staring at the Sun can be dangerous. The only safe way to study the Sun is to clamp the telescope or binoculars firmly in position and project the Sun's image onto a piece of white card. The penalty for attempting to view the Sun directly is blindness.

Sunspot cycles

As astronomers cautiously turned their telescopes on the Sun in the 17th century, they observed that its surface was frequently blemished by variable dark patches, known as sunspots. Most spots are at least the size of the Earth, and some large groups stretch for 150,000 km (93,000 miles) or more – half the distance from the Earth to the Moon. An average spot lasts for a week, although some have been followed for months. By observing the passage of these spots across the Sun's face, astronomers found that the Sun's speed of rotation varies from once every 25 days at the equator to 34 days towards the poles. The fact that the Sun does not rotate uniformly shows that it is not a solid body, but must be composed of incandescent gas.

In 1843, the German astronomer Heinrich Schwabe (1789–1875) found after years of observation that the number of sunspots waxes and wanes in a cycle that is now estimated to last approximately 11 years. At maximum there are dozens of sunspots visible at a time, whereas at minimum there may be no sunspots seen for days on end. The last sunspot maximum occurred in 1979; the next will occur around 1990. The overall storminess of the Sun's surface also varies with the same period, including eruptions known as flares that fling out atomic particles into space and produce radio blackouts on Earth. In fact, in each successive sunspot cycle the north and south magnetic poles of the Sun are reversed, so that a complete cycle of solar activity lasts 22 years.

Why the Sun should have such a variable pulse of activity, and how it affects the climate on Earth, is still unknown; weather patterns on Earth, including rainfall and temperature, seem to change with the sunspot cycle. Looking back through old records, astronomers have found that the solar cycle can be very erratic; from 1640 to 1720 almost no sunspots were seen, and this period coincided with a time of abnormally low temperatures known as the Little Ice Age. Fortunately, we now seem to be in a period of high solar activity and relatively warm temperatures.

A vast powerhouse

What makes the Sun shine? If it were burning like a lump of coal, it would become a dull, burnt-out ember in only a few thousand years. A century ago scientists assumed that the Sun produced its energy by slowly contracting at the undetectable rate of a kilometer or two (about a mile) every 10 years. Such a gradual process would release enough heat to keep the Sun shining for perhaps 100 million years. However, geologists subsequently found that the rocks of Earth are thousands of millions of years old; the most recent evidence from the dating of Earth rocks, Moon rocks, and meteorites has shown that the entire Solar System is 4,600 million years old. To explain what kept the Sun shining for up to one hundred times longer than could be explained by contraction, an entirely new source of power was needed.

The answer came earlier this century with the discovery of the energy contained within the nucleus of the atom. In 1938, the physicist Hans Bethe (b.1906) showed that the center of the Sun is a nuclear powerhouse, turning hydrogen into helium by a series of nuclear reactions, and releasing energy as it does so.

Analysis of light from the Sun shows that it is composed of about 70 per cent hydrogen, the lightest and simplest substance in the Universe, with most of the rest consisting of helium, the second simplest substance. All the other known chemical elements make up only one or two per cent of the Sun. At its center, where temperatures reach 15 million°C and matter is packed 100 times as densely as water, fusion reactions take place like those in a hydrogen bomb. In the reactions, four hydrogen nuclei are fused together to form one helium nucleus. But a helium nucleus is fractionally lighter than four hydrogen nuclei; where does the 'missing mass' go?

The answer is that it is turned into energy, and it is this energy that powers the Sun. Albert Einstein showed in his theory of relativity that matter can be converted into energy, and here in the Sun is the proof. Every second, 600 million tonnes of hydrogen are turned into helium inside the Sun, with 4 million tonnes of hydrogen being converted into energy in the process. The Sun is so massive that, even generating energy at this prodigious rate, it can live for a total of 10,000 million years. It is approximately halfway through its life at present.

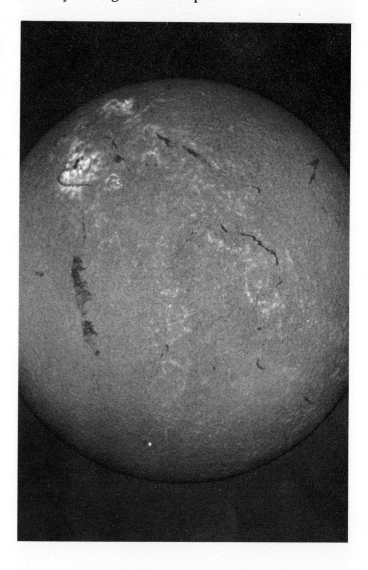

The Sun's layers

Fortunately for us, this fusion reaction proceeds smoothly so the Sun does not fly apart in a bomb-burst. The energy created travels in the form of high-energy radiation such as X-rays for most of the way towards the surface, but completes the journey in gigantic circulating cells of hot gas. At the visible surface of the Sun, called the *photosphere*, the temperature has dropped to 6,000°C.

The photosphere ('sphere of light') is a layer about 300 km (200 miles) thick, and although it looks solid it is actually composed of gas 10,000 times less dense than the Earth's atmosphere. Convection cells of gas about 1,000 km (600 miles) across bubble up through the photosphere like water boiling in a pan, giving a characteristic 'rice-grain' effect to the photosphere – which is where sunspots occur. They are areas of cooler gas (about 4,500°C) that appear dark by contrast with the more brilliant surrounding layers, probably caused by magnetic fields bursting through the surface of the Sun and blocking the outward flow of heat.

Above the photosphere is a layer of gas about 10,000 km (6,000 miles) deep known as the *chromosphere*, meaning color sphere, because of its beautiful pinkish-red color, caused by glowing hydrogen gas. It is visible only at total solar eclipses when the disc of the Moon blocks light from the far more brilliant photosphere. Jets of hot gas known as spicules jut up through the chromosphere, giving its upper edge a jagged appearance.

Occasionally visible at the edge of the Sun are glowing loops of gas called prominences, which stretch away from the Sun's surface like giant arches, thousands of kilometers long. These so-called quiescent prominences are associated with the magnetic fields that loop out of the Sun's surface like strands from a tangled ball of yarn, and can last for weeks or months. Other types, known as surge prominences, eject material at up to 1,000 km (600 miles) per second from flares on the photosphere. The atomic particles ejected in this way reach the Earth, where they cause radio interference and produce beautiful atmospheric displays known as aurorae. All these effects wax and wane with the sunspot cycle.

The outermost layer around the Sun is the *corona*, visible as a faint, pearly-colored halo of light during a total eclipse. It is composed of hot gas boiled off from the Sun and streams outwards into space to form what is known as the solar wind; its atomic particles are detected flowing past the Earth. Therefore we are actually in the outermost regions of the Sun's corona.

Above: Hot gases leap from the Sun's surface to form solar prominences, as photographed from the Skylab space station.

Right: The Sun's corona is shown magnificently in this photograph taken at the total solar eclipse of June 11, 1983.

Left: The stormy surface of the Sun, photographed in the light emitted from hydrogen gas.

Suns in their millions

Stars come in many sizes and brightnesses. Some are much bigger and brighter than the Sun, while others are smaller and fainter. For instance, the red supergiant Betelgeuse is 300 to 400 times the size of the Sun and gives out 15,000 times as much light. The red dwarf Barnard's Star is only about one-tenth the Sun's diameter and gives out one two-thousandth of the Sun's light, so faint that it cannot be seen without a telescope.

Brightness and distance

Although all stars at first glance appear white, closer inspection shows that they are of different colors, which are a guide to their surface temperatures. The hottest stars appear blue or white, whereas the coolest ones appear orange and red. The Sun, which is yellow, is average both in size and temperature. Analysis of a star's light reveals whether it is a hot giant star or a cool dwarf, so that astronomers can calculate how luminous it is. However, stars are at

Above: Dense star clouds towards the center of our Galaxy in the constellation Sagittarius. The dark patches are caused by obscuring dust.

Left: Aldebaran in the constellation of Taurus is a red giant star, about 40 times the diameter of the Sun; here it appears as the smaller bright spot. The larger bright spot is the planet Jupiter.

different distances from us, which affects how bright they appear. Stellar distances are expressed in terms of light years, the distance that light, travelling at 300,000 km (186,000 miles) per second, would cover in a year; a light year is equivalent to 9.5 million million km (6 million million miles). If our Sun were removed to the position of the nearest star, Alpha Centauri, 4.3 light years away, it would appear as the fifth-brightest star in the sky. At a distance of 50 light years, the Sun would be too faint to see without a telescope. The star that appears brightest in the night sky, Sirius, is relatively close – 8.7 light years away. The star Deneb, in Cygnus, gives out 2,000 times as much light as Sirius but it appears fainter to us because it is 200 times farther away. Comparing the calculated value of a star's luminosity with how bright it actually appears in the sky will give you the star's distance.

An important application of this principle is

provided by stars that have a built-in brightness indicator, which is far more accurate than that obtainable by analysis of their light. These are known as Cepheid variable stars, after their prototype, Delta Cephei. Cepheid variables change in brightness every few days as they expand and contract in size, like a slowly beating heart. A Cepheid variable's heartbeat is directly related to its intrinsic brightness, so that by measuring the period of such a heartbeat, astronomers can work out how bright the star is. The diffence between this so-called *absolute magnitude* and the brightness it actually appears from Earth (the *apparent magnitude*) reveals the star's distance; naturally, the farther away from us the star is, the fainter it will appear in the sky. This technique produces much more accurate distance measurements than the method of analysing the star's light described above, but it is hampered by the fact that Cepheid variables are relatively rare – less than 1,000 are known in our Galaxy. Nonetheless, Cepheid variables have proved vital distance indicators in astronomy.

Measuring parallax
The only direct way of measuring a star's distance is by *parallax*, its shift in position as seen from opposite sides of the Earth's orbit. This technique relies on nothing more than simple trigonometry, as in measuring the distance of a tree or building by reference to background objects as seen from two different positions. A star's position is measured relative to background stars, and then is remeasured six months later when the Earth has moved around to the other side of its orbit. The amount of shift reveals the star's distance, the nearest stars showing the greatest parallax. In 1838 the German astronomer Friedrich Bessel (1784–1846) first measured the parallax of a star, 61 Cygni. Bessel calculated that 61 Cygni lay 10 light years away; modern measurements have refined the distance to 11.1 light years, but even so 61 Cygni is still among the 20 nearest stars. Unfortunately, beyond about 100 light years the parallax shift becomes too small to be measured accurately, and astronomers have to rely on the indirect method of brightness comparisons described above.

Novae and other variables
Many stars other than Cepheids vary in brightness. Over 25,000 variable stars have so far been listed, and more are being discovered all the time. Some of them vary regularly, like the Cepheids, but others are much more erratic, notably red giants and supergiants such as Betelgeuse in the constellation of Orion, which are so distended that they become unstable and vary irregularly in size and brightness. As an amateur astronomer, you can make important contributions by monitoring the changes in light output of these stars. The technique compares the brightness of a variable star against nearby stars of known magnitude, to produce a graph of the variable's changes.

Sometimes matter flows between two stars close together, causing sudden eruptions of light. The most spectacular examples of such variables are *novae*, which flare up in the sky where no bright star was seen before. Ancient astronomers believed that they really were new stars; what has really happened is that a formerly faint star has increased by 10,000 times or more in brightness. Novae rise to maximum brightness in a day or two before sinking back into obscurity over days, months, or sometimes years, and are believed to be double star systems in which a white dwarf orbits with another star. Gas from the outer regions of the companion star falls onto the white dwarf, ignites, and is thrown off, causing the sudden surge in brightness. This process can happen more than once, and several novae have been seen to recur after intervals of some years. Amateur astronomers are often the first to spot and report these unexpected stellar eruptions. In a nova explosion only a thin shell of gas is blown off; the star itself does not blow up. By contrast, a *supernova* is the even more brilliant detonation of a massive star which blows itself to bits at the end of its life. The last one in our Galaxy was seen in 1604.

Star groups
Most stars are not single like the Sun, but come in twos, threes, or larger groups. The bright star Castor in the constellation Gemini, for example, is actually a system of six close stars, although only one is visible to the naked eye. A widely spaced double star, Mizar, is the second star along the handle of the Plough. Looking carefully, you will see a fainter star near Mizar; this companion star is called Alcor. A telescope reveals an even fainter star between Mizar and Alcor. Analysis of the spectrum of light from Alcor and this fainter star shows that they both have stars orbiting them, too close to be seen separately; these are known as *spectroscopic binaries*. Alpha Centauri, the closest star to the Sun, is a system of three stars. Binoculars or a telescope reveal that Alpha Centauri itself, which to the naked eye appears as a single brilliant star, is actually double. The third star is a faint red dwarf, called Proxima Centauri, which lies fractionally closer to us than the other two stars. All three stars of Alpha Centauri lie approximately 4.3 light years away.

When one member of a close double star system moves in front of its companion, the star's total light as seen from Earth drops temporarily. Such a system is known as an *eclipsing binary*. A famous example is the star Algol in the constellation Perseus, which drops in brightness every 2.87 days as it is eclipsed by a fainter companion star. The brightness changes of Algol were first explained in 1783 by the English amateur astronomer John Goodricke (1764–1786), from his own naked-eye observations of the star's light. Goodricke was a tragic figure, a deaf-mute who died at the age of 21. But his short life was productive, for in 1784 he discovered another notable eclipsing binary star, Beta Lyrae, which varies in brightness every 12.9 days. These two stars are so close that they are distorted into egg shapes by each other's gravity. Hot gas from the surface of the stars spirals away into space.

Nebulae and the birth of stars
Stars are born from vast clouds of gas and dust known as *nebulae*, the Latin for mist. Nebulae are plentiful throughout the Galaxy: one famous example is the Orion Nebula, visible to the naked eye as a fuzzy patch making up part of Orion's sword. Four new-born stars lie at its heart, and it is the light from the brightest of these stars that makes the nebula shine. Radio astronomers have detected an even larger dark cloud behind the visible bright region of the Orion Nebula. Within this dark region are stars that have not yet switched on. The Orion Nebula, about 15 light years in diameter, is estimated to contain enough material to make a cluster of thousands of stars. Long-exposure photographs through large telescopes reveal other similar clouds that are spawning groups of stars.

Right: Birthplace of the stars. Inside the Orion Nebula, an enormous cloud of gas and dust, new stars are born.

Below: An unusual view of the double star system Alpha Centauri, taken in X-rays by the Einstein Observatory satellite. Alpha Centauri is the nearest star to the Sun, easily visible as a double in small telescopes.

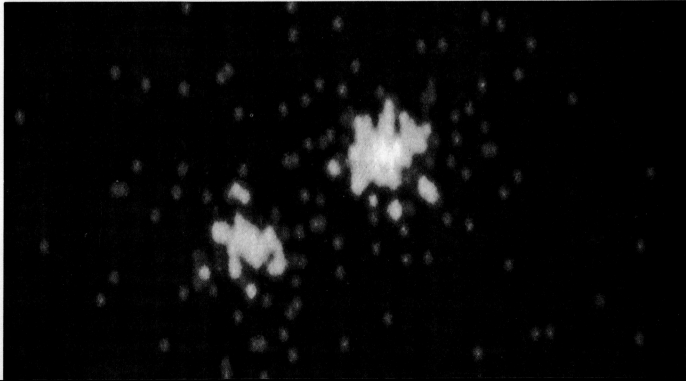

Globule to protostar

Stars begin to form as a cloud breaks up into smaller, denser blobs known as *globules*; these can be seen silhouetted against the brightly glowing background of certain nebulae. A typical globule is about the size of the Solar System, and contains enough mass to make a star the size of the Sun. As the globule shrinks under the inward pull of its own gravity it heats up and begins to glow feebly; it has become a *protostar*. Eventually, perhaps 10 million years or so after the beginning of contraction in the case of our Sun, pressures and temperatures at the protostar's heart have risen sufficiently for nuclear reactions to begin. A star has been born.

If a globule is spinning too fast it will break up as it contracts, thereby producing a close double star; globules that arise close together may remain bound by their mutual gravitational attraction, giving rise to double or multiple stars. Not all of the material from the globule may go into the star, but may remain orbiting the star in a disk, from which a planetary system may form.

Red dwarfs

Calculations by theorists show that for a globule with mass less than about one-tenth that of the Sun, conditions at the center never become extreme enough for nuclear reactions to begin, and so the object never becomes a true star. Stars just above this mass limit are known as red dwarfs; they may be the most abundant stars in the sky and burn their hydrogen fuel so slowly that, despite their small size, they can live far longer than the Sun, up to a million million years. Proxima Centauri is a red dwarf – as is Barnard's Star, the second closest star to the Sun, 6 light years away. Despite their closeness to us, these two stars are so dim that they are invisible without a telescope.

Massive stars

Theorists believe that stars cannot exist with a mass greater than about 100 Suns because they would be unstable and break up. A double star in the constellation Monoceros, known as Plaskett's Star after the Canadian astronomer J.S. Plaskett who first studied it in 1922, consists of two stars each of at least 55 solar masses, making it the most massive pair known. The masses of individual stars are more difficult to calculate, but there is growing evidence to suggest that the theoretical limit on the mass of a star should be revised upwards. One peculiar star in the southern hemisphere, Eta Carinae, is estimated to have a mass of about 200 Suns; it is an erratic variable which gives out as much energy as several million Suns, and is expected to explode as a supernova within the next 10,000 years. Even more extreme is a star known simply as R136a, which lies at the centre of the Tarantula Nebula. Some astronomers think that R136a is a superstar with a mass of 1,000 Suns, giving out the light of 10 million Suns. On the other hand, it could be a small, compact cluster of stars, each with a mass of a few hundred Suns.

Below: The Ring Nebula in Lyra is a shell of gas thrown off by a star (visible at the center of the ring) undergoing its death throes. Our own Sun will end by throwing off a stellar smoke ring like this in billions of years to come.

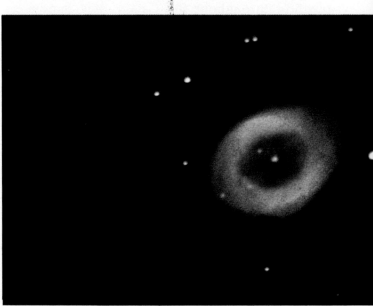

Bottom right: The spider-like shape of the Tarantula Nebula in the Large Magellanic Cloud, a satellite galaxy of our own Milky Way.

Left: The Dumb-bell Nebula is the last gasps of a dying star, visible at the center of the nebula. It lies in the constellation Vulpecula.

Below: The beautiful nebula surrounding Eta Carinae, an unusual star given to sudden fluctuations in brightness. Astronomers expect Eta Carinae to erupt as a supernova.

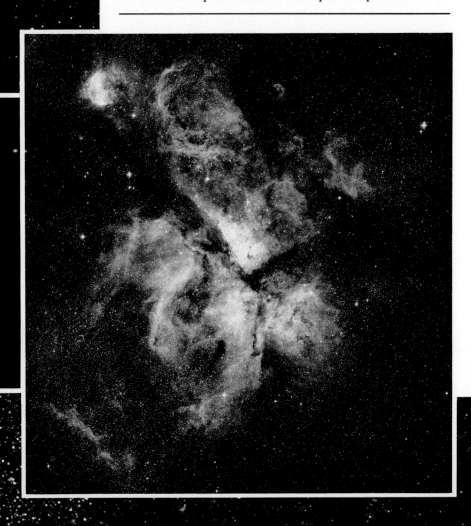

When a large nebula fragments into stars it forms a cluster, such as the group known as the Pleiades in the constellation Taurus. Six or seven of the Pleiades are visible to the naked eye, hence the group's popular name, the Seven Sisters. The youngest members of the Pleiades are no more than one or two million years old; long-exposure photographs show that they are still surrounded by remnants of the cloud from which they formed. When the Sun was born, 4,600 million years ago, it was probably a member of such a cluster which drifted apart over hundreds of millions of years.

Stellar evolution

Nebulae are composed mostly of hydrogen and helium gas, which is also the composition of stars. The conversion of hydrogen into helium by nuclear reactions powers a star for the main part of its life. Currently, the Sun is in stable middle age, about halfway through this hydrogen-burning stage; it will not radically change its output for thousands of millions of years yet.

Eventually a star begins to run out of hydrogen at its center, having turned it all into helium. In search of fresh hydrogen fuel, the nuclear reactions move outwards into the region around the star's core. Here, with more hydrogen to burn, the reactions produce more energy. At the same time, the star's helium core begins to contract, heating up as it does so until it becomes hot enough for the helium nuclei to enter into nuclear reactions of their own, fusing together to form carbon.

With all this extra energy being released at its centre, the star swells up in size, its core becoming hotter as the expanding outer layers turn cooler and redder. It is now a red giant, like the star Aldebaran which forms the glinting red eye of Taurus, the bull. Our Sun will enter this stage about 5,000 million years from now – swelling to 100 times its present size, engulfing the planets Mercury, Venus, and perhaps even Earth. All life will have been extinguished on our planet long before this happens; once the Sun starts to heat up, temperatures on Earth will rise, causing widespread changes in climate, the melting of the polar caps, and consequent flooding of lowland areas. Soon, it will become too hot for any form of life. The seas will evaporate, leaving a parched and barren wilderness. As the Sun swells further, shining 1,000 times as brightly as now, the Earth will be roasted to a cinder in space. Our world will indeed end in fire.

After its brief excursion into super-stardom, our Sun is doomed. Having reached its maximum size, its outer layers will be so distended that they will gently drift off into space, forming a transparent shell of gas similar to the Ring Nebula in Lyra, which resembles a giant smoke ring in space. Such nebulae are called *planetary nebulae* because in small telescopes they look like a planet-like disk.

White dwarf remnants

At the center of a planetary nebula lies the tiny, hot core of the former red giant. This is a white dwarf star, which can contain up to 90 per cent the mass of the original star, packed into a ball no bigger than the Earth. White dwarfs are so dense that a thimbleful of material from one would weigh 10 tonnes, and are so small that they give out only a fraction of the Sun's heat and light. Sirius, the brightest star in the sky, has a white dwarf companion, too faint to be seen without a telescope. This is the remains of a former companion star that evolved more quickly. White dwarfs are bankrupt stars, with no nuclear reactions going on at their cores to produce energy. Over thousands of millions of years a white dwarf cools to a dead ember in space. Thus our once-mighty Sun will end its life as a cold, dark, and invisible ball.

Right: The Pleiades, a cluster of hundreds of new-born stars in the constellation Taurus. The youngest stars are only a few million years old.

Far right: The globular cluster M 15 in the constellation Pegasus is a mass of over 100,000 old stars on the outskirts of our Galaxy.

EXOTIC OBJECTS

The fate of massive stars

How quickly a star evolves depends on its mass. Red dwarf stars age much more slowly than our own Sun, and the largest stars burn out the quickest. If the Sun had 25 per cent more mass, it would already have burned out; a star with ten times the Sun's mass burns out in one hundredth the Sun's lifetime. The largest known stars, such as the two components of Plaskett's Star, live for no more than a few million years.

The early evolution of massive stars is similar to that of the Sun, only much more rapid. Being bigger and hotter than the Sun they at first burn white or blue-white in colour, like Rigel in the constellation Orion, about 900 light years away and 50,000 times more luminous than the Sun. As such stars run out of hydrogen at the center and start to expand, they become red supergiants like Betelgeuse, mentioned earlier, or Antares in Scorpius, 330 light years away and with an estimated diameter 300 times that of the Sun. At this point, their evolution becomes different from that of less massive stars. Whereas stars like the Sun stop at the helium-burning stage, red supergiant stars continue to get hotter and hotter at the core so that a whole range of complex nuclear reactions can take place.

After the fusion of helium at a star's center, a core of carbon is left. The greater gravity of massive stars – those, say, more than ten times heavier than the Sun – squeezes this core until it reaches the critical temperature of 600 million °C, at which carbon fuses to form magnesium, releasing intense heat. Next, as the center of the star gets even hotter, the magnesium enters into reactions to produce other substances; these enter into more reactions, and so on. The process of burning, squeezing, and more burning continues at a runaway rate, building up a mixture of elements at and around the star's core. Outwardly, the supergiant burns bigger and brighter than ever.

Towards a catastrophic end

Inwardly, though, the star is heading for an energy crisis. The day of reckoning comes once its central temperature reaches 3,500 million °C, at which iron nuclei are being formed. Because of iron's atomic structure, the fusion of iron does not release heat but consumes it. With its internal power supply switched off once the fusion of iron begins, the stricken star quickly collapses in upon itself. As the outer layers of the star cascade down upon the core, they ignite in one final explosion. The star has become one of the most violent phenomena in nature – a supernova.

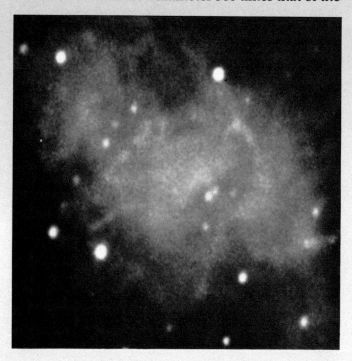

Right: Blown to smithereens – the remains of a star that exploded as a supernova 10,000 years ago, scattered across the sky in the constellation of Vela.

Left: One of the most celebrated objects in the sky is the Crab Nebula, the wreckage of a star Oriental astronomers saw explode in 1054 AD. Near its center, the core of the exploded star remains as a flashing pulsar.

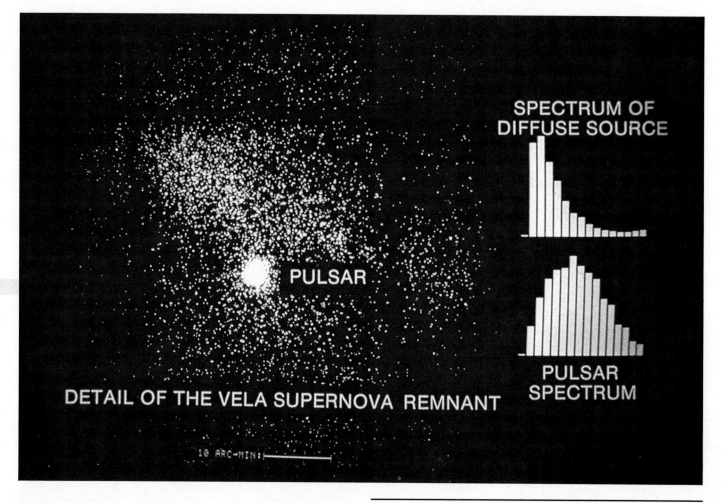

Supernovae

In the nuclear holocaust of a supernova, all the chemical elements in nature are synthesized and scattered into space, where they mix with the existing hydrogen and helium in nebulae, ready to be collected up into new stars and, possibly, planets. The atoms of planet Earth were formed out of the scatterings of ancient supernovae.

The Crab Nebula in Taurus is the shattered remains of a star which was seen by Oriental astronomers to explode in AD 1054. Although the Crab supernova was visible for over 21 months, and at its brightest could be seen in daylight, no western records of it exist, perhaps because European astronomers were still shackled by the Greek doctrine that the heavens were perfect and unchanging and preferred to ignore this embarrassing celestial outburst. The last two supernovae seen in our Galaxy were recorded by Tycho Brahe and Johannes Kepler in 1572 and 1604 respectively. A supernova is estimated to occur about every 50 years in galaxies like ours, so we are long overdue for another one. Until then, astronomers are restricted to observing supernovae in other galaxies.

A supernova brightens over a few days by thousands of millions of times, far more than an ordinary nova. At its brightest, a supernova can rival the combined output of all the stars in a galaxy. Half the star's original mass, or more, may be thrown off at speeds of up to 10,000 km per second (6,000 miles per second). The wisps of gas disperse into space, like the Vela supernova remnant, eventually disappearing from sight after perhaps 100,000 years.

Above: X-ray image of the Vela pulsar, from the Einstein Observatory. Bars at right show distribution of X-rays at different wavelengths from the pulsar, bottom, and the diffuse nebula surrounding it, top.

Top right: Pulsar flashing at the center of the Crab Nebula, imaged in X-rays by the Einstein Observatory satellite.

Bottom right: Photographic sequence of the Vela pulsar, the very faint object near the center of the picture.

Neutron stars

What happens to the core of the star left behind after a supernova explosion? Without the internal energy source of its nuclear fires to sustain it, the core collapses to form a tiny, compressed star even smaller and denser than a white dwarf. The strong inward pull of the heavy core's gravity, aided by the tremendous pressures of the supernova explosion in the layers above it, crush the electrons and protons

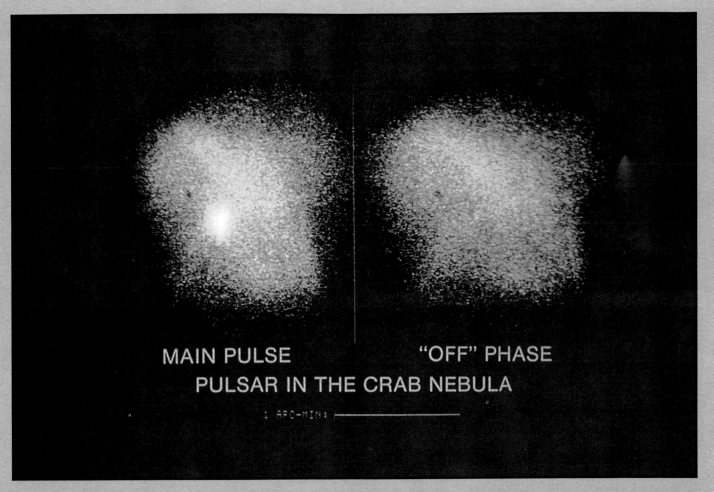

MAIN PULSE "OFF" PHASE
PULSAR IN THE CRAB NEBULA
1 ARC-MIN

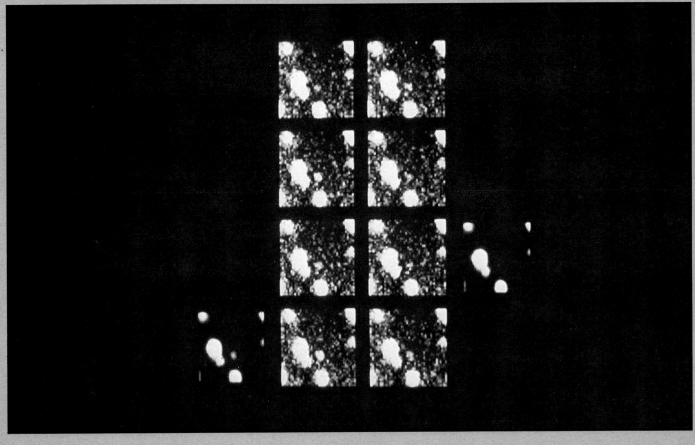

of the core's atoms together to form the electrically neutral particles called neutrons. The resulting object is therefore termed a *neutron star*.

Whereas a white dwarf contains as much matter as the Sun crammed into a ball the size of the Earth, a neutron star has the mass of perhaps two Suns squashed into a sphere no more than 20 km (12 miles) across. The density is such that a thimbleful of neutron star material would weigh an astounding 1,000 million tonnes.

Left: Cygnus X-1 in an artist's impression. Gas from the blue supergiant star is drawn towards the black hole, where it circulates and heats up before plunging out of sight.

Below: Four radio maps at different wavelengths of a supernova remnant known as G 127, that may harbor a black hole.

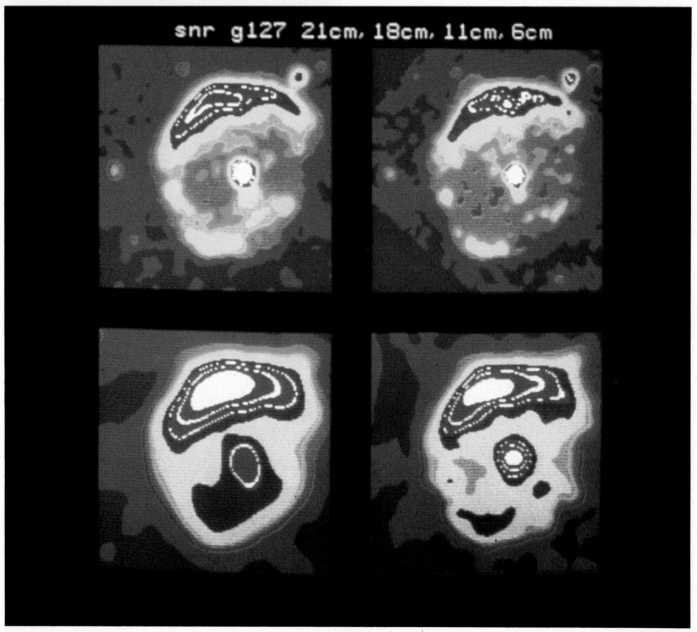

Neutron stars were predicted by theorists as long ago as 1939, but at the time there seemed little chance of detecting them. Then, in 1967, radio astronomers at Cambridge, England, discovered the mysteriously pulsing radio sources which became known as pulsars. The signals were so strange that even transmission by extraterrestrial intelligence was considered – but it soon became clear that these were tiny, rapidly rotating stars that sent out a shaft of energy like a lighthouse beam each time they turned. And the only stars small enough to spin that quickly were neutron stars. How do pulsars pulse? According to the most popular view, the neutron star's strong magnetic field prevents radiation from escaping except at the magnetic poles. Therefore, each time the star's magnetic pole sweeps across our line of sight we see a flash. Pulsars slow down with age, and eventually fade away.

Black holes

Neutron stars are believed to exist in many of the X-ray sources detected by satellites in recent years. These sources are double-star systems in which one star of the pair has evidently already ended its life in a supernova. Gas from the ordinary star plunges into the strong gravitational field of the tiny neutron star, heating up to temperatures of millions of degrees and emitting X-rays.

One X-ray source, Cygnus X-1, contains evidence of an object even more remarkable than a neutron star. The visible companion star of Cygnus X-1 has

Below: Radio picture of SS 433, a curious object that is thought to contain a neutron star or black hole that is ejecting jets of gas, visible here as 'wings'.

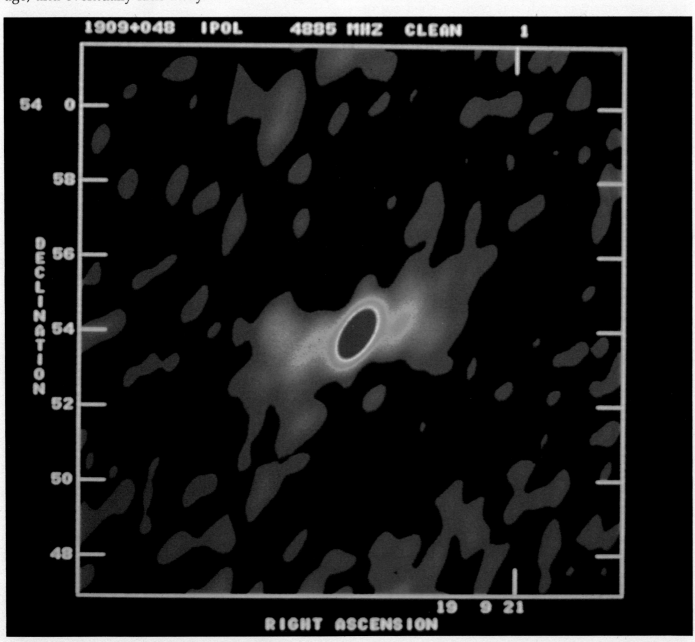

been identified as a blue supergiant star which bears the catalogue number HDE 226868. Observations of this star show that the invisible X-ray source which orbits it has a mass of at least 8 Suns. According to theory, if the core left behind by a supernova weighs more than about 3 solar masses, then its own gravitational pull will compress it even beyond the stage of a neutron star. (Note that here we are talking about the remains of dead stars. While a star is still burning the release of energy from inside prevents it collapsing.) There is no known force that can hold up a dead star weighing more than three Suns against its own gravity. It gets smaller and smaller until eventually it vanishes from sight. It has become a *black hole*.

At the center of a black hole, the original star has been compressed to an infinitely small point, of infinitely high density; in effect, the matter of which the former star was made has been crushed out of existence. Around this central point is a gravitational boundary known as the event horizon from within which nothing can escape, not even light. Thus a black hole is completely invisible. In the case of a 3-solar-mass black hole, the event horizon has a diameter of about 18 km (11 miles), but is larger for greater masses.

Although nothing can get out of a black hole, things can fall in. Imagine the case of a double star system in which one star has already burned out and formed a black hole. Gas will flow from the companion star towards the black hole, heating up and emitting X-rays, as in the case of gas falling onto a neutron star. Although the black hole cannot be seen directly, the X-ray emission gives it away. This is believed to be what is happening in the case of Cygnus X-1.

According to one estimate, there could be as many as 10 million black holes in our Galaxy, formed by the supernova explosions of massive stars. Astronomers are continuing to hunt for more examples of these objects, which are among the most bizarre products of the Universe.

Right: The visible star, arrowed, is a blue supergiant orbited by a black hole. Gas from the visible star falls into the black hole, producing the X-ray emissions that reveal its existence.

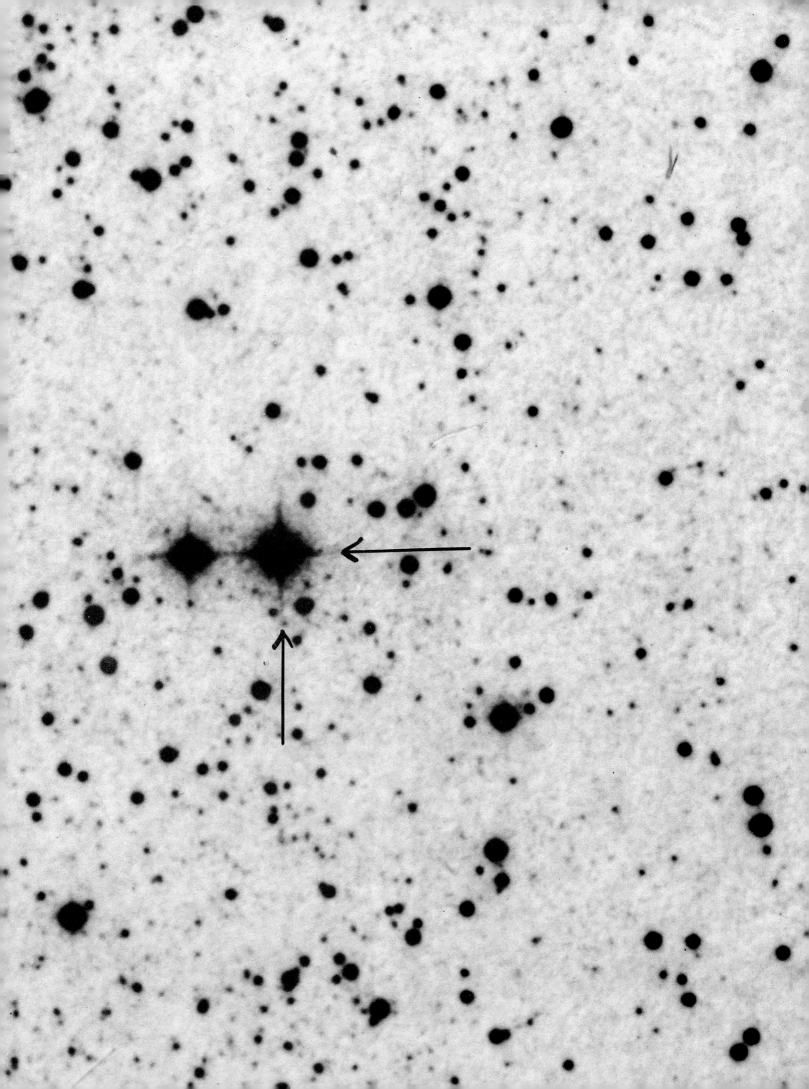

GALAXIES AND BEYOND

No one knows how vast the Universe is. Our concept of the Universe has grown from the cozy, Earth-centered model of the Greeks to a dizzying vision in which the Earth orbits an ordinary star in the outer regions of a Galaxy that consists of a myriad other stars – which is itself only one of countless other galaxies separated by enormous distances – in a Universe of expanding space that may go on without end. The largest telescopes are probing the remotest regions of the Universe in search of further clues to its possible origin.

Until early this century, the Galaxy was presumed to mark the extent of the entire Universe. Then, in 1917, the American astronomer Harlow Shapley showed that the Galaxy was twice as large as previously supposed, and that the Sun was nowhere near the center. He noted that the Sun did not lie symmetrically at the centre of a halo of globular clusters, but instead was shifted to one side – this meant that the Sun could not be centrally placed in the Galaxy. Shapley's results, as refined by later work, demonstrated that the Galaxy is 100,000 light years in diameter, and that the Sun lies 30,000 light years from the center. This was the first accurate measurement of the size of our Galaxy and the Sun's position in it, but it still left open the question of what, if anything, lay *beyond* our Galaxy.

Investigating nebulae
William Herschel had also addressed himself to this problem in the late 18th century when he studied the fuzzy patches known as nebulae which puzzled astronomers: were these part of the Galaxy, or were they beyond it? With Herschel's 1.2-m (48-inch) reflector, then the largest in the world, some nebulae could be resolved into clusters of stars, but others remained obstinately hazy. Certain of these, such as the famous nebula in Orion, were undoubtedly glowing clouds of gas within the Galaxy, but they were the minority.

In Ireland, Lord Rosse (1800–1867) built an even larger telescope than Herschel's to study nebulae. With this 1.8-m (72-inch) reflector, completed in 1845, Rosse found that certain nebulae had a spiral structure – but the significance of his observations

Left: Sir John Herschel, who catalogued the many faint galaxies visible through large telescopes.

Right: Radio map of the Milky Way, color-coded from red (the brightest regions) to blue (the faintest). The radio emissions were caused by electrons spiralling in the Galaxy's magnetic field.

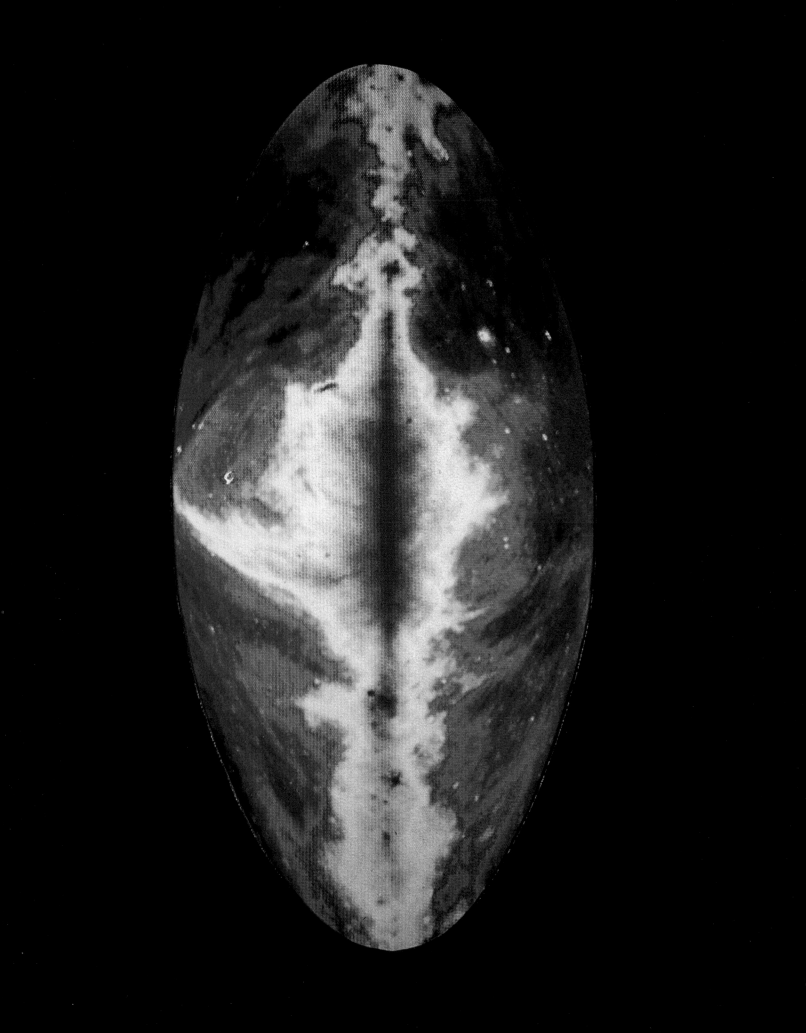

Hubble went on to examine many other galaxies, and in 1925 produced a classification scheme that is still used today. In Hubble's classification, over half of all galaxies are normal Catherine-wheel shaped spirals, like Andromeda and our own Galaxy. These have a bulging heart of old stars, from which emerge curving arms of younger stars. About a quarter of all galaxies are termed barred spirals, because the spiral arms curve from the ends of a straight bar of stars and gas that runs across the galaxy's center. Apart from a few per cent of irregular galaxies which have no obvious shape at all, most other galaxies are classified as ellipticals. These are simply crowded masses of stars near-spherical in shape like a rugby ball. They range from faint dwarf ellipticals, only 5,000 light years or so in diameter containing no more than a million stars, to supergiant ellipticals, the brightest and largest galaxies in the Universe, which can contain 10 million million stars in a diameter of several million light years. Spiral galaxies, by contrast, are more constant: they have between 1,000 million to a

Left: A twin of our own Milky Way, the Andromeda spiral galaxy is the most distant object visible to the naked eye. It lies 2.2 million light years away, so we see it as it appeared 2.2 million years ago.

Right: NGC 1097, a barred spiral galaxy in the constellation Fornax.

Below: M 87 is a supergiant elliptical galaxy in the Virgo cluster that is ejecting a jet of material. M 87 may harbor an immense black hole at its center.

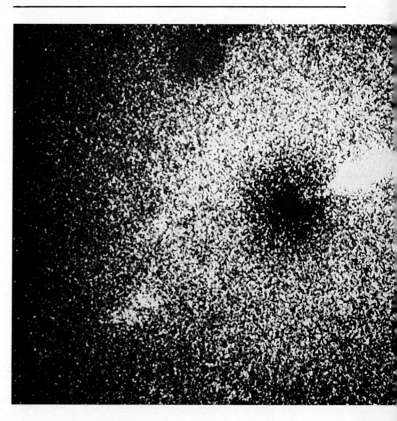

only became clear 75 years later when the American astronomer Edwin Hubble used the newly opened Mount Wilson 2.5-m (100-inch) reflector for a closer look at spiral nebulae.

Hubble's discovery of galaxies

In 1924, Hubble photographed faint stars in the outer regions of the so-called Andromeda Nebula, a well-known spiral, which showed that it was actually a separate star system, or galaxy, similar to our own. By implication, therefore, *all* the thousands of spiral nebulae then known to astronomers were actually separate galaxies. This discovery opened up the vision of an awesomely vast Universe with galaxies scattered like islands throughout a sea of space. Today thousands of millions of galaxies are within range of the largest telescopes. On a clear night, the Andromeda Galaxy is visible as a faint hazy patch even without binoculars; it is the farthest that the naked eye can see. According to modern figures, the Andromeda Galaxy is 2.2 million light years away, so we see it as it was when our ape-ancestors roamed the plains of Africa. Yet it is still one of our closest galactic neighbors.

Above: Radio galaxy Centaurus A appears to be a giant elliptical galaxy encircled by a lane of dark dust. It may have resulted from the merger of an elliptical galaxy and a spiral galaxy.

million million stars, and their diameters range from 20,000 light years to several hundred thousand light years.

Our Galaxy

The spiral structure of our Galaxy has been traced in detail by radio astronomers, who pick up radiation at 21 cm wavelength from the hydrogen gas that lies plentifully between the stars in the Galaxy's spiral arms. It is from this gas that new stars form. Our Galaxy is estimated to contain at least 100,000 million stars, making it one of the largest spirals. The faint band of stars that crosses the sky, the Milky Way, is the rest of our Galaxy as seen from inside; the name Milky Way is often used as a synonym for our Galaxy. Our Sun takes about 225 million years to orbit the Galaxy's center, so that it has gone around only about twenty times since it was born. All the stars are moving as the Galaxy rotates. Their movements, known as *proper motions*, are too slight to be noticeable to the naked eye in a human lifetime, but over many thousands of years they will gradually change the shapes of the constellations.

The Magellanic Clouds and the Local Group
Two irregular-shaped galaxies known as the Magellanic Clouds accompany our Milky Way like satellites. Named for the round-the-world explorer Ferdinand Magellan, they are visible as faintly luminous clouds in the southern hemisphere. The larger of the two lies about 180,000 light years away and contains approximately 10,000 million stars, about one-tenth the number in our Galaxy. The smaller Magellanic Cloud, with about 2,000 million stars, lies 230,000 light years away. The Clouds are 40,000 and 30,000 light years in diameter, respectively, and are our closest neighbors in a vast cluster of about 30 galaxies known as the Local Group. Most galaxies throughout the Universe seem to cluster in groups, some containing as many as several thousand members.

Right: Star field in Sagittarius, looking towards the center of our Galaxy. The long streak is the trail of a satellite.

Below: The Large Magellanic Cloud is a satellite galaxy of our own Milky Way.

The expanding Universe

Our Local Group is bound together by its own gravitation. As Edwin Hubble looked deeper into space with the Mount Wilson telescope, he found that other galaxies appeared to be moving away from us at speeds that increased with their distance. The galaxies' movements could be told from the so-called *red shift* of their light, a lengthening of wavelength caused by recession; since red wavelengths are longer than blue ones, a lengthening of wavelength means a reddening of light, hence the term red shift.

Another name for this change of wavelength is the Doppler effect, after the German physicist Christian Doppler (1803–1853); it is also used to detect the motions of stars. The red shift of light from the galaxies told Hubble that they were receding, and as he painstakingly measured the distances to the galaxies, he found that the farthest ones were receding the fastest: the entire Universe seemed to be expanding. Hubble's sensational discovery, announced in 1929, is a keystone of modern cosmology – the branch of astronomy concerned with the origin and evolution of the Universe.

This expansion does not mean that our Galaxy is the center of the Universe. As the Universe expands, each galaxy moves away from the others because the space between them has grown. An observer in any one galaxy will see the same effect as an observer in any other galaxy. Therefore there is no center to the Universe. First Shapley, by dethroning the Sun from the centre of the Galaxy, and then Hubble, with his demonstration that the Galaxy is as helpless as an unmoored ship against the tide of the Universe, had wrought a revolution even more astounding than that in which Galileo

---

Right: The celebrated Whirlpool Galaxy, M 51, was the first galaxy in which spiral structure was noticed.

Below left: An artist's impression of our local Supercluster of galaxies. Most important is the Virgo cluster (left), with thousands of galaxies.

Below: M 81, a beautiful spiral galaxy lying 10 million light years away in the constellation Ursa Major.

and Kepler affirmed the theory of Copernicus, three centuries earlier.

Big Bang versus Steady State
Knowledge that the Universe is expanding led the Belgian astronomer Georges Lemaître (1894–1966) to suggest that the Universe began in a giant explosion, now known as the Big Bang; the galaxies are pieces from that explosion flying outwards. Modern measurements of the expansion of the Universe show that the galaxies are receding at a rate of about 24 km (15 miles) per second for every million light years of distance; this figure is known as *Hubble's constant* and it is important because it relates a galaxy's distance to its speed of movement. Once the red shift in a galaxy's light has been measured, therefore, astronomers can deduce its distance from Hubble's constant. Hubble's constant also reveals the age of the Universe, because by tracking back the rate of expansion astronomers can deduce when the Big Bang explosion must have taken place. Current estimates suggest that about 15,000 million years has elapsed since the Big Bang.

In 1948, three astronomers working in Britain, Thomas Gold (b.1920), Hermann Bondi (b.1919) and Fred Hoyle (b.1915) put forward a very different, controversial theory of the Universe – the Steady State theory. They proposed that the Universe did not have a beginning but has always existed. Instead of there being a single instant of creation, new matter continually comes into being to fill the space left as the Universe expands, so that if we could see the Universe at any time in the past or future it would look much the same as it does today.

However, observations have indicated that the Universe was *not* the same in the past. Although it seems impossible to turn back the clock to see what the Universe looked like, this is exactly what astronomers do by looking far off into space. Because light takes a finite time to reach us, we see objects in the Universe not as they appear now but as they were when the light left them. For instance, we see the Sun as it was 8.3 minutes ago, Sirius as it was 8.7 years ago, and the Andromeda Galaxy as it was 2.2 million years ago. Telescopes are time machines; they allow astronomers to see the most remote areas of the Universe as they appeared thousands of millions of years ago.

Right: A curving bridge of hydrogen gas links the spiral galaxy M 81, at right, with a small neighbor galaxy, NGC 3077, at left, as revealed in this radio map.

Below: Doppler effect. When a galaxy is receding from us, lines in its spectrum are shifted towards the red end of the spectrum: its light is redshifted (center). If the galaxy is approaching, its light is blueshifted (bottom). Compare with the spectrum of light from a galaxy that is not moving relative to the Earth (top).

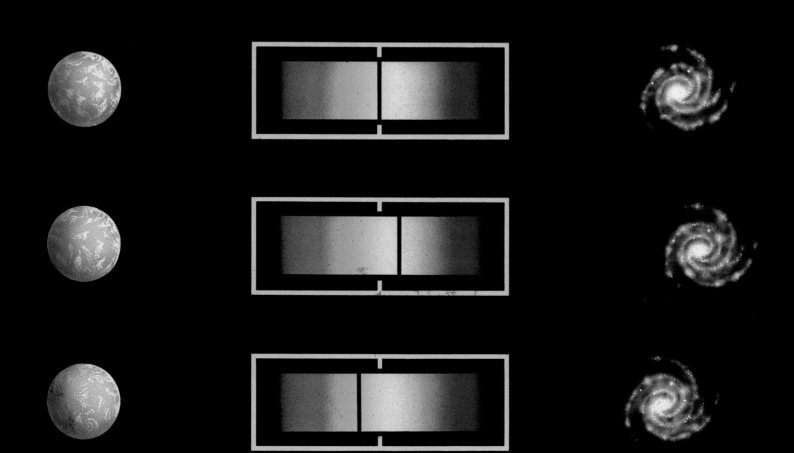

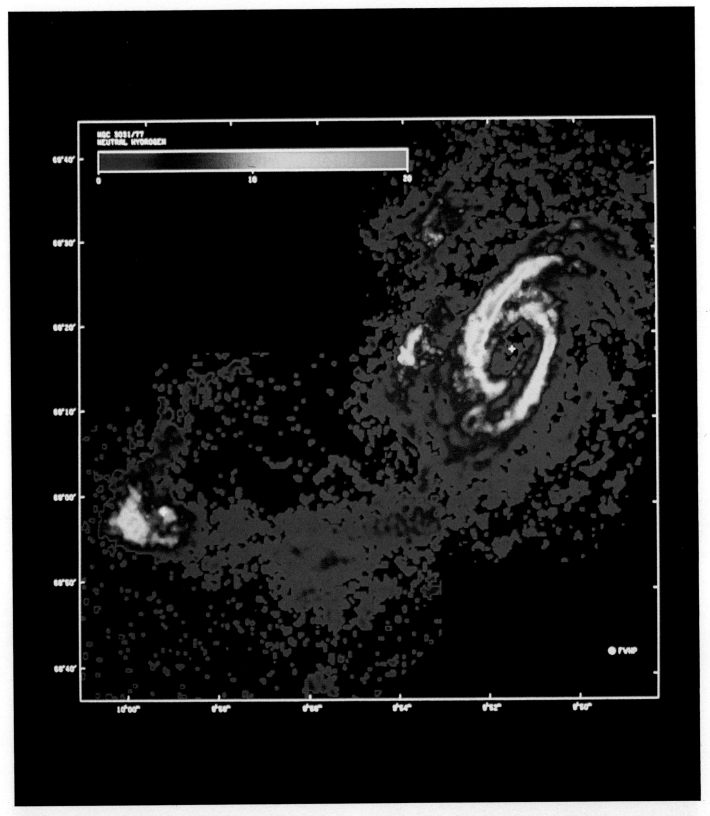

The farthest known galaxies are receding at about half the speed of light, which places them around 8,000 million light years away; therefore we see them as they appeared long before the Earth was born. Beyond this, galaxies are too faint to be seen by optical telescopes on Earth, but they can still be detected by radio astronomers, who during the 1950s began counting the number of radio sources deep in the Universe. They found that the weakest sources, which were presumably the most remote, were more numerous than the Steady State theory accounted for. It seemed as though the Universe looked different in the past, which would mean it was evolving, as the Big Bang theory required.

Rapidly receding quasars

Bigger trouble for the Steady State theory came as astronomers began to identify certain of the radio sources optically. Some of them were clearly associated with known galaxies, albeit peculiar-looking ones, but others were centered on what looked like faint blue stars. In 1963, the American astronomer Maarten Schmidt (b.1929), examining the light from one of these objects known as 3C 273 (its number in the third catalogue of radio sources compiled at Cambridge, England) found that it had a red shift so great that it must be far off in the Universe, and was therefore not a star at all. The red shift of 3C 273 was 16 per cent the speed of light, meaning that it was moving at 48,000 km (30,000 miles) per second; from the Hubble constant, it must be 2,000 million light years away. At the time, this was an unprecedented distance, but soon other star-like objects with even greater red shifts were identified. They became known as quasi-stellar objects, shortened to *quasars*.

Quasars have proved to be the most astounding objects in the Universe. Less than one light year in diameter, they pack as much energy as thousands of galaxies combined, so that they are visible over vast distances at which normal galaxies cannot be seen. The farthest known quasars are receding at over 90 per cent the speed of light, which means that they lie about 14,000 million light years away, and hence

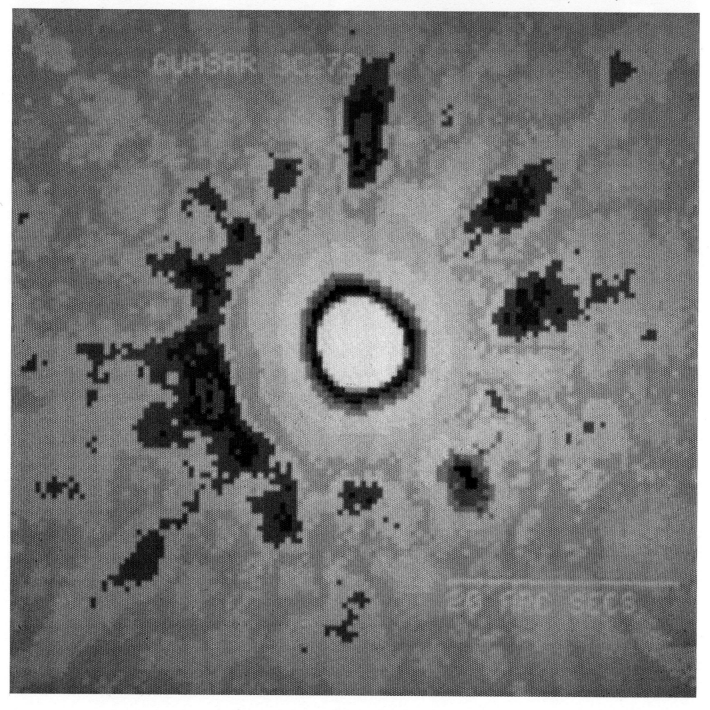

we see them as they appeared shortly after the formation of the Universe.

Quasars present astronomers with an energy crisis, because no one is certain how so much luminosity can be produced in such a relatively tiny space. Some astronomers began to wonder in desperation whether the red shift of quasars really was produced by their motion at all, or whether it had some other cause which would mean that they were not so distant or luminous as otherwise assumed. However, many independent lines of investigation have confirmed the conventional view that quasars really are far out, and since no such objects are observed nearby, the Universe must have looked

Left: The quasar 3C 273 taken at X-ray wavelengths by the Einstein Observatory satellite. The blob at the four o'clock position is part of a jet of gas apparently being ejected from the quasar at high speed.

Below: A double quasar as seen from Earth. At top is the direct view of the quasar, with radio lobes either side of it. Center bottom is the second image of the quasar's core, with the intervening galaxy just above it.

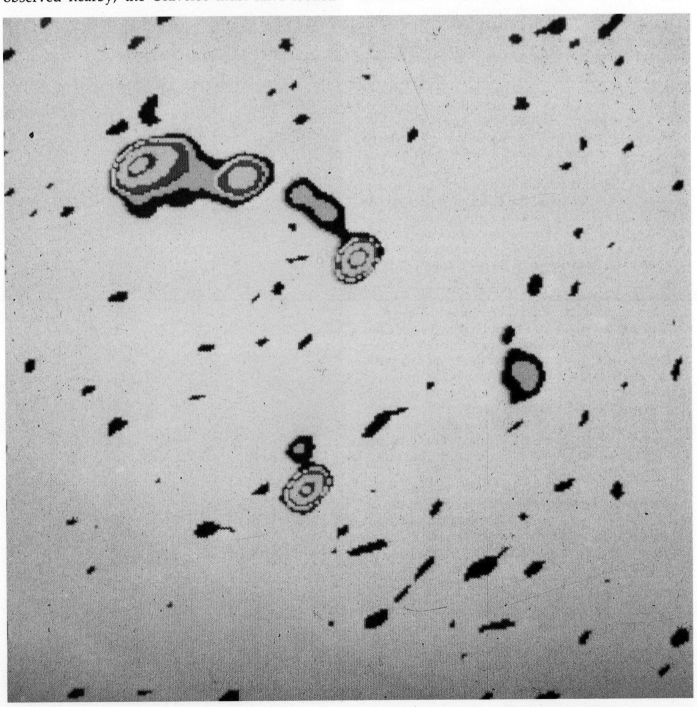

very different in the past – in contradiction of the Steady State theory.

To confuse matters further, not all quasars emit radio waves. Those that do have radio emissions from two optically invisible clouds, one on either side of the visible quasar. This double structure is the same as that shown by radio galaxies, which are visible galaxies that strongly emit radio waves. One famous example is the strange-looking galaxy known as NGC 5128, its number in the *New General Catalogue of Nebulae and Clusters* published in 1888 by the Danish astronomer J.L.E.Dreyer (1852–1926). This object looks like a supergiant elliptical galaxy ringed by a band of dark dust. Radio-emitting regions flank it, as though ejected in one or more immense explosions in the past.

Radio galaxies do not give out as much energy as quasars, probably because they have quietened down with age. Another clue to the nature of quasars comes from spiral galaxies with brilliant nuclei, known as Seyfert galaxies after their discoverer Carl Seyfert (1911–1960), an American astronomer. These seem to be a midway stage between quasars and normal galaxies. If Seyfert galaxies were moved farther away, only their brilliant centers would be visible, and they would look like quasars. In fact, some quasars have been found to be surrounded by a fuzz, like the faint outer regions of a galaxy. Therefore most astronomers now believe that quasars are the extremely active centers of galaxies at an early stage in their evolution.

Steady State debunked
A coherent picture is at last beginning to emerge. If current theories of stellar evolution are correct, giant black holes would be expected to form at the centers of galaxies as a result of the death of massive stars. These black holes would grow by the influx of material to a mass of a million Suns or so, at which stage gravity would tear apart nearby stars and then suck in the gas. Such a massive black hole swallowing matter at the heart of a galaxy could be the energy source that powers quasars, ejecting radio-emitting clouds.

Astronomers thinking along these lines have tacitly accepted the Big Bang theory. However, the major body blow to Steady State came not from quasars, but from the observation by radio astronomers, Arnio Penzias and Robert Wilson, in 1965. They found that the Universe seems to have a slight background warmth – that it is not totally cold, but

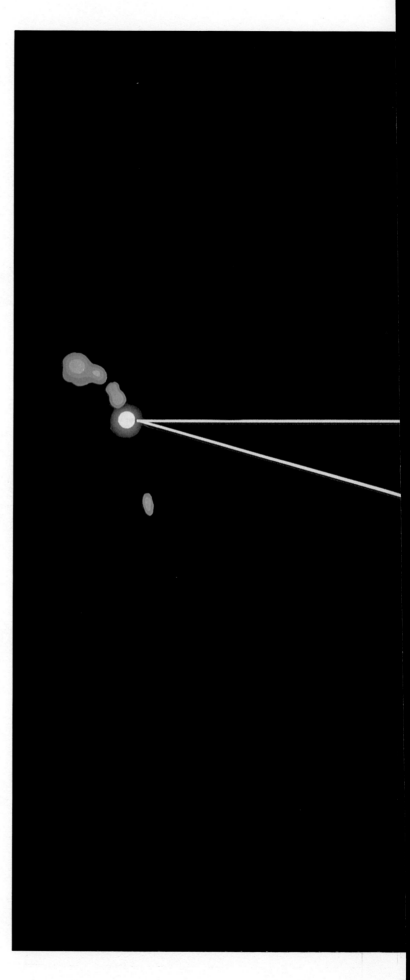

Right: A double image of one quasar formed by the bending of its light around an intervening galaxy. Its light and radio waves not only travel directly to us, forming one image, but its radiation is also bent by the gravitational field of a foreground galaxy, forming a second image.

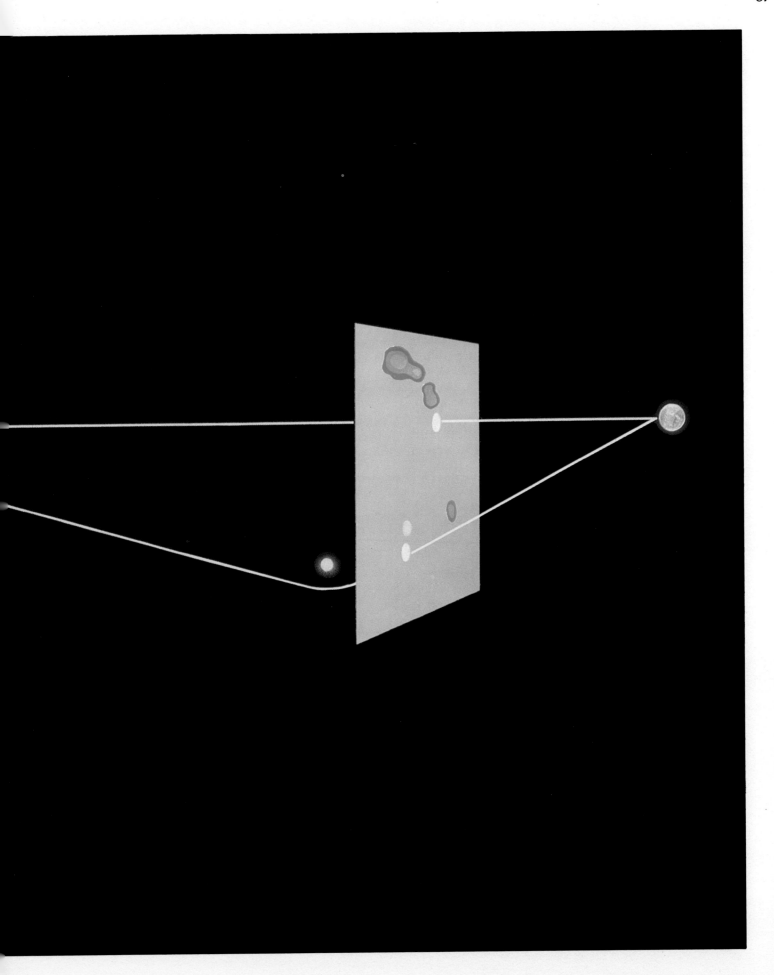

has a temperature of 2.7°C above absolute zero (absolute zero is the coldest temperature possible, equivalent to −273°C). This cosmic background radiation is interpreted as being heat left over from the Big Bang; its existence had actually been predicted as long ago as 1948 by the American astronomer George Gamow (1904–1968). With its discovery, all remaining support for the Steady State theory dwindled.

Cosmologists now sketch the following life story of the Universe and of the galaxies in it. Fifteen thousand million years ago, or thereabouts, all the matter of the Universe was compressed in a superdense state, from which it exploded in the Big Bang. Hydrogen and helium gas flung out from the Big Bang broke up over thousands of millions of years into clumps which shrank to form galaxies.

Our Galaxy began to form about 14,000 million years ago, which is the age of the oldest stars in the halo of globular clusters around the Galaxy. These formed first as that massive parent gas cloud shrank, while a nucleus of stars built up at the center. Around this nucleus the remaining gas was spread out into a disk by the Galaxy's rotation, producing the spiral arms in which younger stars were born. Then, 4,600 million years ago, our Sun and its planetary system came into being in one of those spiral arms. Now, intelligent beings sit on the third planet from the Sun and ask questions about their environment in space.

We shall continue to ask questions, for it would be foolish to think that we have the final answers to such fundamental questions as the nature and origin of the Universe. For instance, it seems that the remaining gas and dust in our Galaxy, and others, will one day all be collected up into stars, with no more left to make new stars. Meanwhile the Universe will continue to expand, so that the galaxies will thin out in space, eventually losing sight of each other. Over millions of millions of years the Universe must slowly fade out and die.

But in, say, another 50 years our ideas may be completely different. In centuries to come, our present views of the Universe may seem as naive as those of the ancient Greeks do to us today.

Right: M 77, a spiral galaxy in Cetus, is a Seyfert galaxy; it has an unusually small, bright center.

Below: The center of Seyfert galaxy NGC 4151 may harbor a supermassive black hole with a mass of 1,000 million suns.

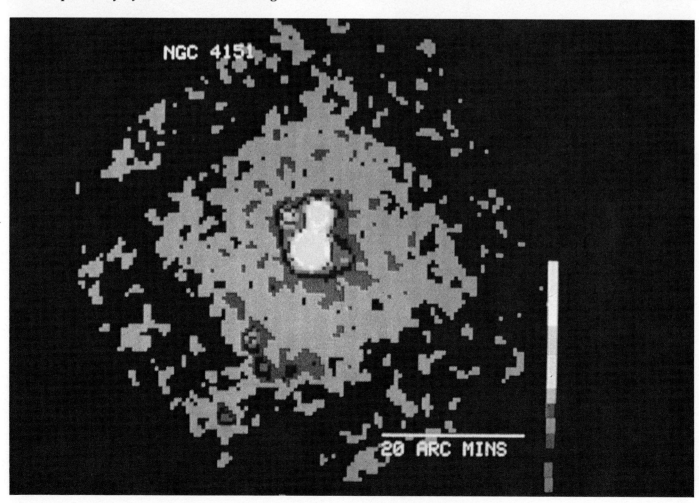

THE NEW ASTRONOMY

Objects in space emit not only visible light. They also give out radiation at longer and shorter wavelengths, from radio waves to X-rays, that are invisible to the eye. Since World War II, astronomers have been opening up new windows on the heavens to study these other wavelengths. Major advances have come through radio astronomy, which detects long-wavelength radiation in space. An American radio engineer, Karl Jansky (1905–1950), accidentally detected the first radio noise from the Galaxy in 1932 while searching for sources of noise on radio-telephone transmissions. Jansky's discovery was followed up by the radio ham Grote Reber (b.1911) but it was not until after the War that radio astronomy began to flourish.

Most radio telescopes work rather like reflecting telescopes. A large reflector panel collects radiation and focuses it to a point where it is detected and amplified electronically. The output from radio telescopes is usually stored on magnetic tape. Because radio waves are so much longer than light waves, radio telescopes have to be much bigger than optical telescopes to see the sky in as much detail. In some cases, radio astronomers electronically combine the signal from a line of individual dishes to synthesize

Below: Individual dishes comprising the Very Large Array radio telescope in New Mexico.

the view of the sky that would be seen by one enormous radio dish.

The largest fully steerable radio dish in the world is the 100-m (328-ft) telescope of the Effelsberg radio observatory near Bonn, West Germany. The largest single dish in the world is 305 metres (1,000 ft) in diameter at Arecibo, Puerto Rico, slung in a natural hollow between the mountains. It cannot be steered, but scans the sky overhead as the Earth spins. On the plains near Socorro, New Mexico, lies the Very Large Array, completed in 1980. This remarkable instrument consists of 27 antennae arranged along Y-shaped arms up to 21 km (13 miles) long. Their output, when combined, gives a view of the sky equivalent to that from a single dish 27 km (17 miles) in diameter.

The Space Telescope
The best view of the sky is, however, obtained by observing it from space without the blurring effect of the Earth's atmosphere. With the launch of the Hubble Space Telescope by the Shuttle next year, distant objects will be observed with ten times the clarity previously achieved by existing telescopes on Earth. With regular servicing by astronauts, the

Top: The striking Sombrero galaxy, a spiral galaxy with a dark lane of dust, seen almost edge-on and thus named because of its resemblance to a Mexican hat.

Top left: The world's largest fully steerable radio telescope, at Effelsberg, West Germany.

Left: The movable dishes of the Very Large Array equal the performance of a single dish 27 km (17 miles) in diameter.

Above: Artist's impression of the Hubble Space Telescope in orbit. It is the largest telescope ever put into orbit, and should create a revolution in astronomy.

Left: Honeycomb structure makes the 2.4-m (94-inch) diameter mirror of the Hubble Space Telescope as lightweight as possible.

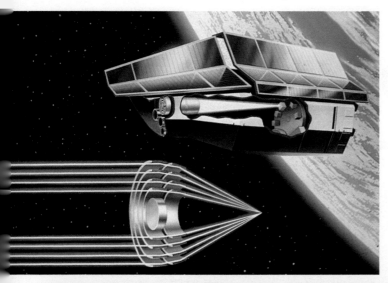

Far left: The Infra Red Astronomy Satellite, IRAS, during checkout before its launch on January 26, 1983.

Left: X-rays are so energetic that they must be bounced at a grazing angle off cylindrical mirrors which focus them onto a detector, as shown here in the Einstein Observatory satellite.

Below: One of the new wavelengths opened up by satellites is the infrared. This is an infrared view of the Andromeda galaxy from IRAS, the Infra Red Astronomy Satellite.

Bottom: Compare this view of the Orion Nebula at radio wavelengths with the traditional optical view on page 61.

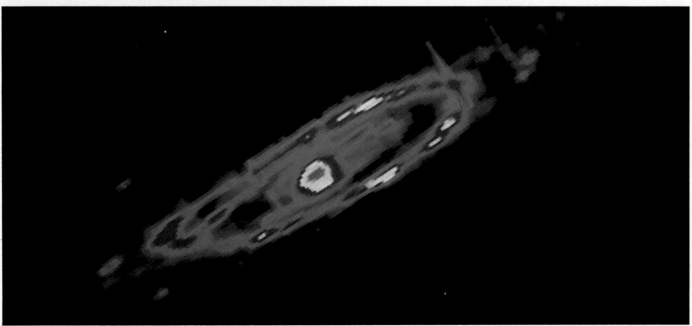

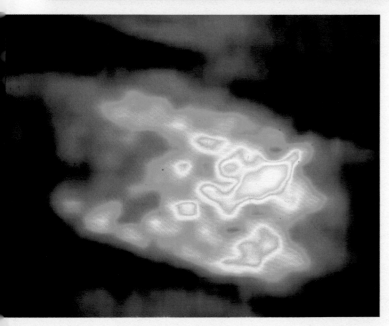

Space Telescope should remain operational for ten years or more.

There are many wavelengths of radiation that can be studied only from orbit, because they do not penetrate the atmosphere at all. These wavelengths include infra-red, ultraviolet, X-rays, and gamma rays. Satellites have been sent up to scan the skies at these wavelengths, telling astronomers about processes including the birth and death of stars, and violent events such as super-hot gas pouring into black holes in space. The opening up of these new windows on the Universe is producing a revolution in astronomy similar to that which followed the invention of the telescope. Astronomers themselves are now able to travel into orbit on the Space Shuttle to conduct their own observations with instruments mounted in the Shuttle's cargo bay. What they will find cannot be foreseen; that is essentially the excitement of the discovery of our Universe.

INDEX

Page numbers in *italics* are for illustrations

Adams, John Couch 48
Alcor 60
Aldebaran *58–9*, 64
Algol 60–1
Almagest, the 10
Alpha Centauri 59, 60, *61*
Anaximander 10
Andromeda Galaxy 76, *76*, 80, 82, *95*
Andromeda Nebula 76
Anglo-Australian Observatory, New South Wales 21
Antares 66
Apollo space mission 36
Aristarchus 10
Aristotle 10
asteroids 50, *50*
aurorae 57

Barnard's Star 58, 62
Bessel, Friedrich 60
Beta Lyrae 61
Beta Pictoris 49, *49*
Betelgeuse *26*, 58, 60, 66
Bethe, Hans 56
Big Bang theory 82, 83, 86, 88
binoculars 22
black hole 71–2, 86
Bondi, Hermann 82
Brahe, Tycho 12, 68

Callisto 44
Cassini, Jean Dominique 46
Castor 60
Centaurus A radio galaxy *78*, 80
Cepheid variable stars 60
Ceres 50
Cerro Tololo, Chile 21
Charon 49, *49*
Columbus 10
comets 31, 51
constellations 8
Copernicus, Nicolaus 10, 11–12, *14*
cosmology 80
Crab Nebula *67*, 68, *69*
Cygnus X-1 *70*, 71–2

Deimos 41
Delta Cephei 60
Deneb 59
dish *see* telescope, radio
Doppler, Christian 80
Doppler effect 80, *82*
Dreyer, J. L. E. 86
Dumb-bell Nebula *63*
dwarf stars, red 62, 66
dwarf stars, white 64, 70

Earth 10, 11, 31, 52, *52*, 53, 54, 56, 64, 68
eclipse *see* Sun; Moon
eclipsing binary 60
Effelsberg radio observatory, Bonn, West Germany 92, *92*
Einstein, Albert 56
Einstein Observatory satellite *95*
ellipses *12*
energy 56–7
Eratosthenes 10

Eta Carinae 62, *63*
Europa 44, *45*
European Southern Observatory, Chile 21
galaxies 76–8, 83, 88
  barred spiral 76, *77*
  elliptical *76*, 76–7
  Local Group 79, 80, *81*
  radio 86
  Seyfert 86, *88*, *89*
  spiral 76, *76*, 86, *93*
Galaxy 21, 74, 78–9, 88
Galilei, Galileo 12, *12*, 14, 44
Galle, Johann Gottfried 48
Gamow, George 88
Ganymede 44
Geminids 52
geocentric theory 8
globules 62
Gold, Thomas 82
Goodricke, John 61
Gregory, James 15

Hale, George Ellery 18
Hale reflector, California 21
Hall, Asaph 41
Halley, Edmond 51
  comet 51
HDE 226868 72
heliocentric theory 10, 11–12, 14–15
Hermes 50
Herschel, Sir John 74
Herschel, Sir William 16, 47, 74
Hipparchus 10
Horsehead Nebula *2–3*
Hoyle, Fred 82
Hubble, Edwin 21, 76, 80
Hubble's constant 82
Hubble Space Telescope 92, *93*

Infra Red Astronomy Satellite (IRAS) 49, *94*
Io 44, *44–5*

Jaipur, India *9*
Jansky, Karl 91
Jet Propulsion Laboratory, Pasadena, California 45
Jodrell Bank, England 7
Jupiter 14, 31, *42–3*, 42–5, *58–9*

Kepler, Johannes 12, 68
Kettering School, England 8
Kitt Peak Observatory, Arizona *1*, *4–5*, 16, 21, *21*

Lemaitre, Georges 82
Leonid meteor shower *52–3*
Leverrier, Urbain 48
Lick Observatory, California 18, *18*
light year 59
Lilge, Al 28
Lippershey, Hans 12
Lyra, Ring Nebula *62–3*, 64
Lowell, Percival 38, 48
Luna 3 space probe 36

M 15 *64–5*

M 77 89
M 81 *80*, *83*
M 87 *76–7*
Magellan, Ferdinand 79
Magellanic Clouds 79, *79*
Mariner space missions 32, 40
Mars 11, *30*, 31, 38–41, *40–1*
Mauna Kea, Hawaii 20, 21
Mercury 11, 31–2, *32*, *33*
meteors 52, *52–3*
Meteor Crater, Arizona 53
meteor showers 52, *52–3*
meteorites 30, 52
Milky Way 14, 16, *19*, 75, 78, 80
Mimas *46–7*
Mizar 60
Moon 10, 14, *17*, 24, 34–7, *35*, *36*
  eclipse 10, 35
Mount Wilson reflector, California 18, 21, 76
Multiple Mirror Telescope, Mount Hopkins, Arizona 21

nebulae 61, *61*, 62–3, 64, 74–6
Neptune 16, 31, 48, *48*
Nereid 48
neutron stars 69–71
Newton, Sir Isaac *14*, 15
NGC 1097 *77*
NGC 3077 *83*
NGC 4151 *88*
NGC 5128 86
novae 60
Nut *8*

Orion *8*, *26*
Orion Nebula *26*, 61, *61*, *95*

parallax 60
Penzias Arnio 86
Perseids 52
Phobos 40, 41
photography 22, 28
Piazzi, Giuseppe 50
Pioneer-Venus orbiter 32, 33
planets, creation of 30
Plaskett, J. S. 62
Plaskett's Star 62, 66
Pleiades 64, *64*
Pluto 16, 48–9, *49*
protostar 62
Proxima Centauri 60, 62
Ptolemy, Claudius 10, *11*
pulsars 68, 69, *69*, 71
Pythagoras 10

quasars 84, *84*, *85*, 86, *86–7*

R 136a 62
radiation wavelengths 91
radio astronomy 91
Reber, Grote 91
red giants 60, 64
red shift 80
Rigel *26*, 66
Rosse, Lord 74

Sagittarius 59, *79*
Saturn 11, 16, 31, 45–7, *46*, *47*
Schiaparelli, Giovanni 38

Schmidt, Maarten 84
Schwabe, Heinrich 56
Seven Sisters *see* Pleiades
Seyfert, Carl 86
Shapley, Harlow 18–21, 74, 80
shooting stars *see* meteors
Sirius 59, 64, 82
61 Cygni 60
Solar System 30, 56
Sombrero galaxy *93*
spectroscopic binaries 60
spectroscopy 18
spiral arms *76*, 88
SS 433 71
star patterns *see* constellations
star clouds *59*
stars 58–73
  groups of 60–1
Steady State theory 82, 84, 86
Stonehenge, England 6, *6*
Sun 10, 21, 30, 31, 54–7, *54*, 56, 64, 66, 74, 78, 82, 88
  corona 29, *53*, 57, *57*
  eclipse 10, 25, 28, *29*, 57
  prominences 57, *57*
sundials *9*
sunspots 55–6, *55*, 57
supergiants 60
  blue *70*, 72, *72–3*
  red 66
supernova 60, 66, *66*, 67–9, *70*

Tarantula Nebula *62–3*
telescope mountings 27
telescopes *13*, 17–18, 25
  multi-mirror 21
  radio 90–1, *91*, *92*, *93*
  reflector 15, *15*, *16*, *18*, 18–21, *23*
  refractor *12*, *14*, 18, *23*
  Schmidt *23*
3C 273 84, *84*
Titan 46
Tombaugh, Clyde 48
Triton 48
Tycho's supernova 12

Universe 8, 80–3, 88
Uranus 16, *17*, 31, 47–8

Vela 66, 68, *68*, *69*
Venus 11, 14, 31, 32–4, *33*
Venus space missions 34
Very Large Array radio telescope, New Mexico 90–1, *92*, *92*
Viking space missions 40–1
volcanoes 31, *31*, 40
Voyager space probes 44, 46

Whirlpool Galaxy (M 51) *80*
Wilson, Robert 86

X-ray sources 71–2

Yerkes Observatory 18

Zelenchukskaya reflector, USSR 21

PICTURE CREDITS

Broadhurst Clarkson and Fuller 17 bottom, 18 top **Daily Telegraph Photo Library** 70 top **Michael Holford** 8, 15 **The Mansell Collection** 11-14 **The Photo Source** 90-91 **Ian Ridpath** title, 19, 44, 62-63 bottom, 63, 65, 77, 78, 80 right **Royal Observatory, Edinburgh** 61 top **The Science Photo Library Limited** endpaper, 6, 16, 17 top, 20-43, 45-59, 61 bottom, 62-63 top, 64-65, 66-69 bottom, 70 bottom, 71-77, 79, 80 left, 82-89, 92, 93 bottom left and right, 94, 95 **Tony Stone Associates** 7 **Zefa** half title, contents, 9, 18 bottom.

Front Cover: **Tony Stone**
Back Cover: **Zefa**

Multimedia Publications (UK) Limited have endeavored to observe the legal requirements with regard to the rights of suppliers of photographic material.